T0344450

Internet of Things (IoT) Enabled Automation in Agriculture

About the Authors

Dr. Rajesh Singh is currently associated with University of Petroleum and Energy studies, Dehradun as Associate Professor and with additional responsibility as Head Institute of Robotics Technology (R&D). He has been awarded as gold medalist in M.Tech. and honors in his B.Tech. His area of expertise includes embedded systems, robotics and wireless sensor networks. He has organized and conducted a number of workshops, summer internships and expert lectures for students as well as faculty. He has twelve patents in his account. He has published around hundred research papers in referred journals/conferences.

Under his mentorship students have participated in national/international competitions including Texas competition in Delhi and Laureate award of excellence in robotics engineering in Spain. Twice in last four years he has been awarded with certificate of appreciation from University of Petroleum and Energy Studies for exemplary work. He got Certificate of appreciation for mentoring the projects submitted to Texas Instruments Innovation challenge India design contest, from Texas Instruments, in 2015.He has been honored with young investigator award at the International Conference on Science and Information in 2012. He has published books on "Embedded System based on Atmega Microcontroller" with NAROSA publishing house and "Arduino based Embedded System" with CRC/Taylor and Francis. He is editor to a special issue published by AISC book series, Springer with Title Intelligent Communication, Control and Devices-2016 & 2017.

Anita Gehlot has more than ten years of teaching experience with area of expertise embedded systems and wireless sensor networks. She has ten patents in her account. She has published more than fifty research papers in referred journals and conference. She has organized a number of workshops, summer internships and expert lectures for students. She has been awarded with certificate of appreciation from University of Petroleum and Energy Studies for exemplary work. She has published books on "Embedded system based on Atmega microcontroller" with NAROSA publication house and "Arduino based Embedded System" with CRC/Taylor and Francis.She is editor to a special issue published by AISC book series, Springer with Title Intelligent Communication, Control and Devices-2017.

Bhupendra Singh is Managing director of Schematics Microelectronics and provides Product design and R&D support to industries and Universities. He has completed BCA, PGDCA, M.Sc. (CS), M.Tech and has more than eleven years of experience in the field of computer networking and Embedded systems.He has published books on "Embedded System based on Atmega Microcontroller" with NAROSA publishing house and "Arduino based Embedded System" with CRC/Taylor and Francis.

Dr. S. Choudhury is the Head of theDepartment of Electronics, Instrumentation, and Control in the University of Petroleum and Energy Studies. He has teaching experience of 26 years and he has completed his Ph.D from the University of Petroleum and Energy Studies, M.Tech (Gold Medalist) from Tezpur Central University, Tezpur, India and received his B.E. degree from NIT, Silchar University, India. He has published more than 70 papers in various national/international conferences/journals. He has filed ten patents. His area of interest is Zigbee-based wirelessnetworks. He has been selected as the outstanding scientist of the twenty-first century by the Cambridge Biographical Centre, UK. He has also been selected in the who's who of the world in science by Marquis Who's Who, USA.He has published books on "Embedded System based on Atmega Microcontroller" with NAROSA publishing house and "Arduino based Embedded System" with CRC/Taylor and Francis. He is editor to a special issue published by AISC book series, Springer with Title Intelligent Communication, Control and Devices-2016 & 2017.

Internet of Things (IoT) Enabled Automation in Agricultural

Rajesh Singh
Anita Gehlot
Bhupendra Singh
Sushabhan Choudhury

CRC Press is an imprint of the
Taylor & Francis Group, an **informa** business

NEW INDIA PUBLISHING AGENCY
Pitam Pura, New Delhi – 110 088

2023

Milton Park, Abingdon, Oxon, OX14 4RN

ess
und Parkway NW, Suite 300, Boca Raton, FL 33487-2742

dia Publishing Agency

ı imprint of Informa UK Limited

Cataloguing-in-Publication Data
ord for this book is available from the British Library

428741 (hbk)
428765 (pbk)
364702 (ebk)

781003364702

es New Roman

Preface

The primary objective of writing this book is to provide a platform to beginners to get started with Internet of things based'Embedded system'with basic knowledge of the programming and interfacing of the devices.

The book comprises of eleven chapters including introduction of IoT, NodeMCU and interfacing of input output devices. The book is intend to serve for the students of B.Tech/B.E, M.tech/M.E, Ph.D scholars and who needs the basic knowledge to develop real time system using NodeMCU.

We acknowledge the support from www.nuttyengineers.com for using its hardware products to demonstrate and explain the working of the systems.We would like to thank the publisher for encouraging our idea about this book and the support to manage the project efficiently.

We are grateful to the honorable Chancellor Dr. S.J Chopra, Mr. Utpal Ghosh (President & CEO, UPES), Dr. Deependra Jha (Vice Chancellor, UPES), Dr. Kamal Bansal (Dean, CoES, UPES), Dr. Piyush Kuchhal (Associate Dean, UPES) and Dr. Suresh Kumar (Associate Dean, UPES) for their support and constant encouragement. In addition we are thankful to our family, friends, relatives, colleagues and students for their moral support and blessings.

Although the circuits and programs mentioned in the text are tested on real hardware but in case of any mistake we extend our sincere apologies. Any suggestions to improve in the contents of book are always welcome and will be appreciated and acknowledged.

Rajesh Singh
Anita Gehlot
Bhupendra Singh
Sushabhan Choudhry

Contents

1

Introduction to IoT

To understand IoT "Internet of Things" it is important to understand internet first. Internet is connecting a device to other device anywhere in the world. When two devices are connected with the Internet, they can send and receive all kinds of information such as text, graphics, voice, and video. The high-speed, fiber-optic cables are used, through which the bulk of the Internet data travels. The Internet has revolutionized the communication world. In today's scenario the internet is an important part for people around the world. The Internet is a global network of billions of electronic devices. The World Wide Web (www) is most widely used part of the Internet.

The "Internet of Things is a potentially integrated part of the 'Future Internet'. IoT can be defined as a dynamic global network with self - configuring capabilities based on communication protocols where physical and virtual 'Things' interact with each other.

The Internet of Things (IoT) is inter-connecting the networks of smart devices. Smart devices may be embedded with vehicles, buildings, or other electronics object with software, sensors, actuators, and network connectivity. Network connectivity enables smart devices to collect and exchange data. The Internet of Things (IoT), also referred to as the Internet of Everything (IoE) which comprises of the web-enabled devices. IoT has its applications in smart city, smart surveillance, automated transportation, smarter energy management systems, water distribution, urban security and environmental monitoring. Agriculture is another new area of IoT application.

Due to advancement in technology the processing power and storage capacity has been increased, in the recent years. At the same time technology is making the devices pervasive, wearable and mobile. Smart devices are fitted with sensors, actuators which enable devices to sense, compute and act intelligently. They are able to exchange information with environment.

IoT is new era of research and challenges for the researchers. Imagine some smart devices which can sense environment parameters and only can be accessed by authenticate people on mobile phone or laptop. Mobile App can be developed to access or control the smart devices. The major features of IoT includes collection and transmission of data, triggering the actuators corresponding to the sensory data received and communication with other members of network.

Interaction with the Internet

The ability of communication makes the IoT devices different from the ordinary sensor devices. A sensor generates data which needs to be managed and the embedded memory is limited, so people look for alternative solutions like storing data on memory cards. In IoT devices are connected to network to store the information online. By this method the data can be stored online as well as can be accessed anytime from anywhere in the world. The actuators can be triggered by the data received form the environment. The data management and exchange of information is based on cloud computing. Fig.1 illustrates the IoT network and other services.

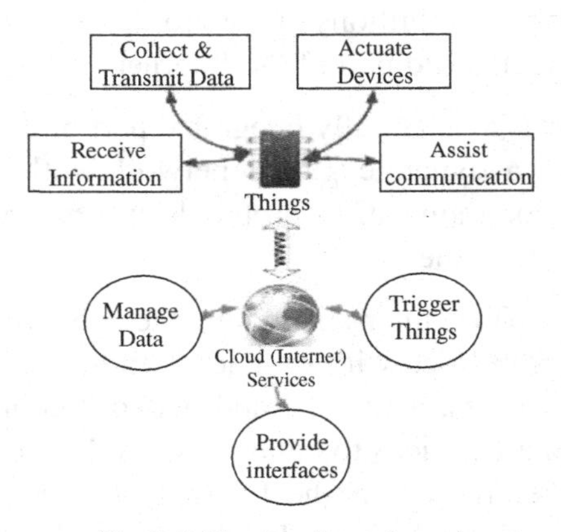

Fig.1: IoT networks and services

Major components of "Internet of Things" device include control unit, sensor, power source and communication modules. The evolution of IoT depends on the technical progress of controllers and wireless modules with efficient power. The platform for IoT enabled devices are specially designed with internet web protocols which allows easy communication.

Examples of IoT services include data logger like ThingSpeak and the iDigi Device Cloud. User can visualize and manage data with these services. It is feasible to connect microcontroller platforms like Arduino directly to

IoT services. Small wearable devices with IoT communication with power efficiency is the target of the market. Imagine the IoT enabled devices which can automate the agriculture field and generate alerts with minimum intervention of human being, this is the target of this book. Explore the book for real time sensory data acquisition and forecasting with appropriate examples. IoT application in the agriculture field with development of smart devices is future of smart farming.

2

Role of Ardiuino in Agricultural Field

2.1 Introduction

Automation of agricultural field with IoT and microcontroller is new era of technology. Arduino based solution to agricultural field is open source platform. It can be used for developing data loggers and new sensors environment. The rate of solutions with Arduino and sensors is high as it is open source platform. The use of Arduino in agriculture can accelerate the growth of automation. Arduino can be interfaced with sensors like:

- Temperature
- Humidity
- Soil Moisture
- Water Pressure
- Water Flow and Water Metering
- Rain Gauge
- Wind Speed

Fig.2.1 shows the block diagram of sensors interfacing with Arduino. The DS180 sensor, Ultrasonic sensor, Current sensor and voltage sensor are connected as digital out, serial out, analog out and analog out respectively. All the sensors are power up by +5V power supply.

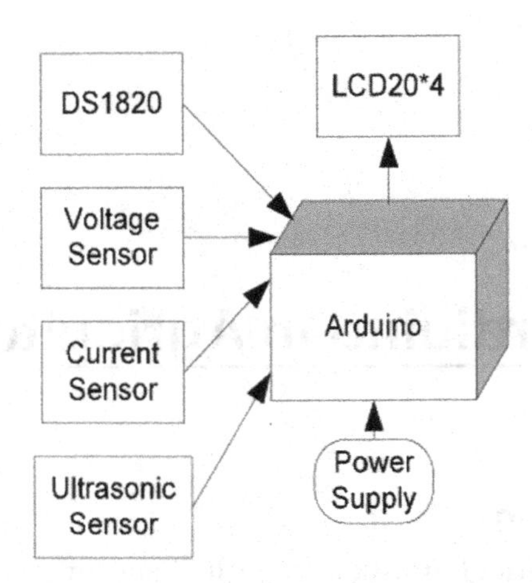

Fig.2.1: Block diagram to interface Arduino with sensors

Table:2.1 Component list

Component/Specification	Quantity
Power supply12V/1A	1
Arduino development board	1
Jumper wire M-M	20
Jumper wire M-F	20
Jumper wire F-F	20
DS18S20	1
Voltage sensor DC	1
Current sensor DC	1
Ultrasonic sensor	1
LCD20*4	1
LCD patch/explorer board	1
LM35	1

Note: All components are available at www.nuttyengineer.com

2.2 Circuit Diagram

Connections

1. RS pin of LCD is connected to pin12 of Arduino Uno
2. RW pin of LCD is connected to GND pin of Arduino Uno
3. RS pin of LCD is connected to pin11 of Arduino Uno

4. D4 pin of LCD is connected to pin5 of Arduino Uno
5. D5 pin of LCD is connected to pin4 of Arduino Uno
6. D6 pin of LCD is connected to pin3 of Arduino Uno
7. D7 pin of LCD is connected to pin2 of Arduino Uno
8. Pins 1and 16 should be connected to GND pin of power supply patch.
9. Pins 2 and 15 should be connected to +5V pin of power supply patch.
10. Variable lag of 10K POT should be connected to pin 3 of LCD
11. +12V power supply jack is connected to DC jack of Arduino Uno.
12. Connect +Vcc and GND pins of voltage sensor to the +5V and GND pins of power supply patch
13. OUT pin of voltage sensor is connected to A0 pin of Arduino UNO
14. Connect +Vcc and GND pins of Current sensor to the +5V and GND pins of power supply patch
15. OUT pin of current sensor is connected to A1 pin of Arduino UNO
16. Connect +Vcc and GND pins of DS1820 sensor to the +5V and GND pins of power supply patch
17. OUT pin of voltage sensor is connected to 8 pin of Arduino UNO
18. Connect +Vcc and GND pins of Ultrasonic sensor to the +5V and GND pins of power supply patch
19. OUT pin of voltage sensor is connected to RX(0) pin of Arduino UNO

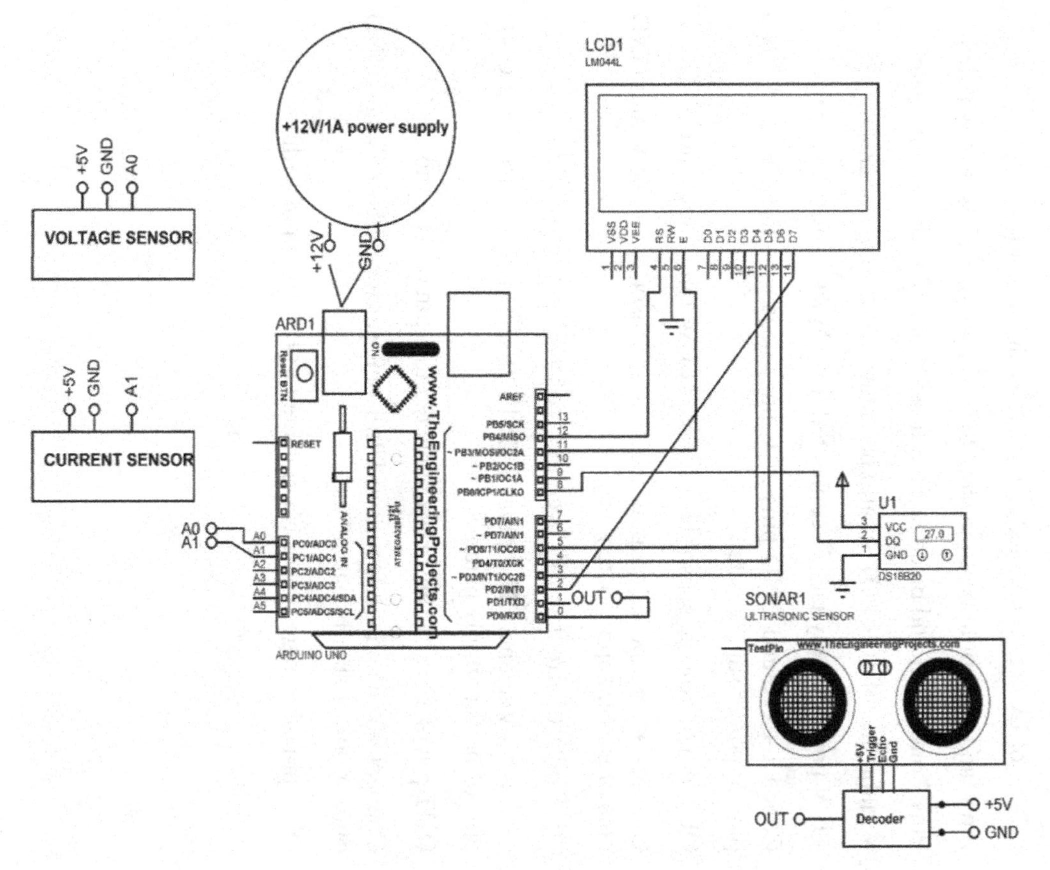

Fig.2.2: Circuit Diagram

2.3 Program

```
#include <OneWire.h>
#include <DallasTemperature.h>
#include <LiquidCrystal.h>
LiquidCrystal lcd(12,11,5,4, 3, 2);
#include <SoftwareSerial.h>
SoftwareSerial rajSerial(6,7);// 6 rx /7 tx
////////////////////////// for DS16B20
#define ONE_WIRE_BUS 8
OneWire oneWire(ONE_WIRE_BUS);
DallasTemperature sensors(&oneWire);
/////////////// for Ultrasonic sensor
String inputString_ULTRA = “”;        // a string to hold incoming data
boolean stringComplete = false;
/////////voltage measurement
int VOLTAGE_SENSOR_PIN=A0;
float OUTPUT_DC = 0.0;
float INPUT_DC_V = 0.0;
float R1 = 30000.0; // vlaue of R1 resistor
float R2 = 6565.0; // vlaue of R1 resistor
int V_LEVEL = 0;
/////////current measurement
const int  CURRENT_SENSOR_PIN=A1; //Connect current sensor with A1
of Arduino
int mVperAmp = 66; // use 100 for 20A Module and 66 for 30A Module
int I_LEVEL = 0;
int ACSoffset = 2430;
double Voltage = 0; //voltage measuring
double INPUT_DC_I = 0;// Current measuring
```

```
String ULTRA;
void setup(void)
{
 // start serial port
 Serial.begin(115200);
 mySerial.begin(9600);
 inputString_ULTRA.reserve(200);
 // Start up the library
 sensors.begin();
 lcd.begin(20, 4);
 lcd.setCursor(0,0);
 lcd.print("TRANSFORMER HEALTH");
}
 void loop(void)
{
  /// measure ultrasonic sensor
      ULTRASONIC_READ();
 /////// measure voltage
      V_LEVEL = analogRead(VOLTAGE_SENSOR_PIN);
      OUTPUT_DC = (V_LEVEL  * 5.0) / 1024.0; // see text
      INPUT_DC_V = OUTPUT_DC / (R2/(R1+R2));
////////////// measure current.............
      I_LEVEL = analogRead(CURRENT_SENSOR_PIN);//reading the value
from the analog pin
      Voltage = (I_LEVEL  / 1024.0) * 5000; // Gets you mV
      INPUT_DC_I = ((Voltage - ACSoffset) / mVperAmp);
sensors.requestTemperatures();
int TEMP=sensors.getTempCByIndex(0)
```

```
//////////////send serial
Serial.print(TEMP);
Serial.print(",");
Serial.print(ULTRA);
Serial.print(",");
Serial.print(vin);
Serial.print(",");
Serial.print(Amps);
Serial.print('\n');
///////////display on LCD
lcd.setCursor(0,1);
lcd.print("TEMP:");
lcd.setCursor(5,1);
lcd.print(TEMP);
lcd.setCursor(0,2);
lcd.print("Oil-Level:");
lcd.setCursor(12,2);
lcd.print(ULTRA);
lcd.setCursor(0,3);
lcd.print("V:");
lcd.setCursor(3,3);
lcd.print(INPUT_DC_V);
lcd.setCursor(10,3);
lcd.print("I:");
lcd.setCursor(13,3);
lcd.print(INPUT_DC_I);
delay(20);
```

```
}
void ULTRASONIC_READ()
{
 while (rajSerial.available()>0)
{
inputString_ULTRA = rajSerial.readStringUntil('\r');// Get serial input
ULTRA=String(((inputString_ULTRA[0]-48)*100) + ((inputString_ULRA[1]
-48)*10)+((inputString_ULTRA[2]-48)*1))+"."+String(((inputString_
LTRA[4]-
48)*10)+((inputString_ULTRA[5]-48)*1));
}
inputString_ULTRA  = "";
delay(20);
}
```

3

Role of NodeMCU in Agricultural Field

3.1 Introduction

Internet of Things (IoT) allows computational devices and sensory support to connect with each other and access services on the Internet. The IoT idea was introduced to connect devices through the Internet and facilitate access to information for users. The potential application of IoT includes agriculture. The aim of this chapter is to explain the role of NodeMCU in the agriculture field. NodeMCU is IoT module which can be used to present the IoT concept as a basis for monitoring and control. The systems used in farm production processes. NodeMCU play a key role, with a focus on their realization by available microcontroller platforms and appropriate sensors. IoT based system provides opportunity to users to monitor and control the process remotely.

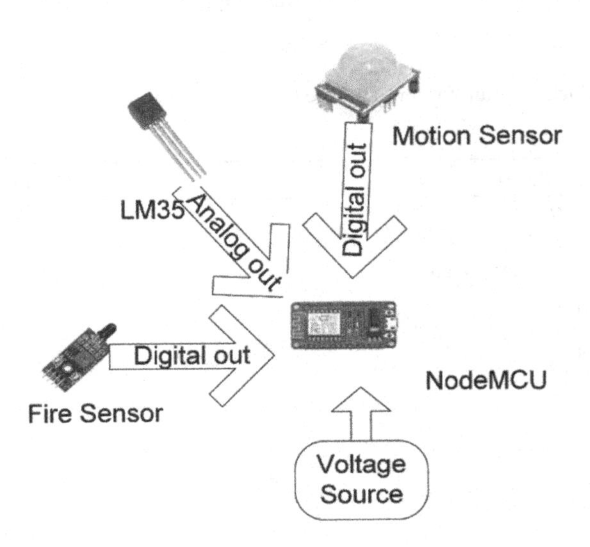

Fig.3.1: Generalized block diagram

Fig.3.1 shows the generalized block diagram of sensors (analog and digital) interfacing with NodeMCU. The sensor LM35 temperature sensor, Fire sensor and motion sensor are connected with node MCU via analog out, digital out and digital out mode respectively. All the sensors are powered up by +5V power supply.

3.2 Temperature sensor interfacing with NodeMCU

To understand the interfacing of sensor with NodeMCU a simple temperature sensor interfacing is described in this section. Fig.3.2 shows the block diagram of temperature sensor (LM35) interfacing with NodeMCU. It comprises of NodeMCU, LM35, LCD and power supply.

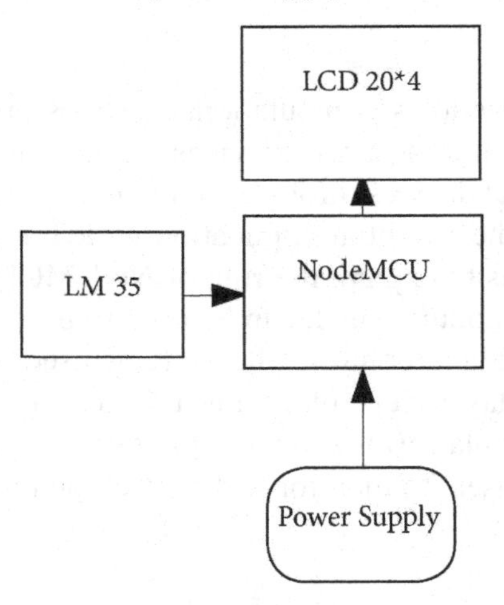

Fig.3.2 Block diagram for temperature sensor interfacing with Node MCU

Table 3.1: Component List

Component/Specification	Quantity
Power supply 5V/1A	1
NodeMCU development board	1
Jumper wire M-M	20
Jumper wire M-F	20
Jumper wire F-F	20
Patch for Node MCU	1
LCD20*4	1
LCD patch/explorer board	1
LM35	1

Note: All components are available at www.nuttyengineer.com

3.2.1 Circuit Diagram

Connections

1. NodeMCU D0pin is attached with RS pin of LCD
2. RW pin of LCD is connected to Ground
3. NodeMCU D1 pin is attached with E pin of LCD
4. NodeMCU D2 pin is attached with D4 pin of LCD
5. NodeMCU D3 is attached with D5pin of LCD
6. NodeMCU D4 pin is attached with D6 pin of LCD
7. NodeMCU D5 pin is attached with D7 pin of LCD
8. Pin 1 and pin 16 of LCD is connected with Ground
9. Pin 2 and pin 15 of LCD is connected with +Vcc

Fig.3.3 shows the circuit diagram of the system.

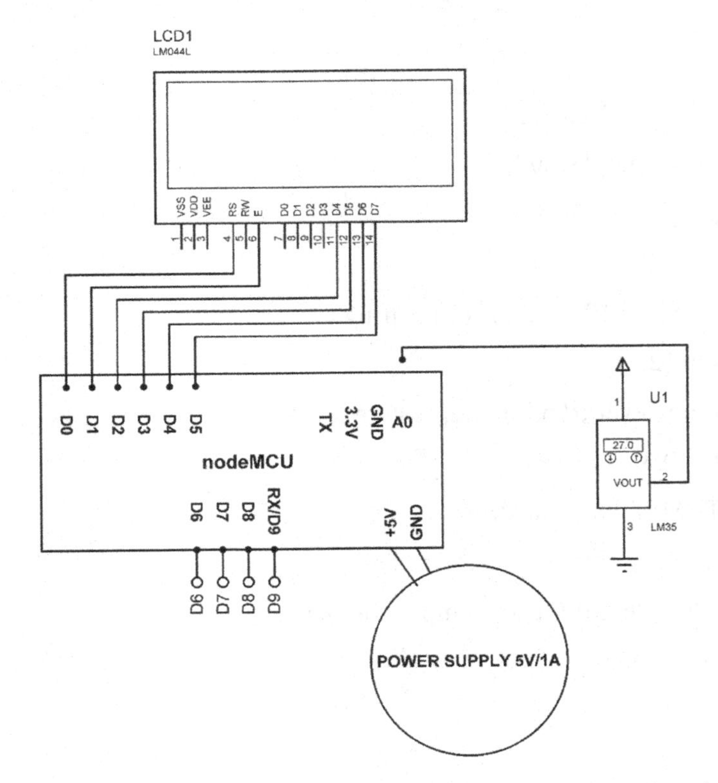

Fig.3.3: Circuit diagram of the temperature sensor interfacing with NodeMCU

3.2.2 Program

```
#include <LiquidCrystal.h>
LiquidCrystal lcd(D0, D1, D2, D3,D4, D5);//// RS RW E D4 D5 D6 D7
int Temp_level=0;
int Actual_temp=0;
void setup()
{
lcd.begin(20, 4);///// lcd initialise
lcd.setCursor(0,0);
lcd.print("Temp Monitoring SYS");
delay(2000);
}
void loop()
{
Temp_level=analogRead(A0);
Actual_temp=Temp_level/2;
lcd.clear();
lcd.setCursor(0,2);
lcd.print("TEMP_LEVEL:");// print string
lcd.setCursor(12,2);
lcd.print(Temp_level);// print temperature level
lcd.setCursor(0,3);
lcd.print("TEMP:");// print string
lcd.setCursor(6,3);
lcd.print(Actual_temp);// print temperature level
delay(1000);
}
```

4

Development of BLYNK app to Automate Agriculture Field

4.1 Introduction

The objective of this chapter is to design a mobile app for agricultural field automation with Blynk app. Mobile app is designed to control the water management in field as per requirement. The complete system comprises of two sections sensor node and mobile app. Sensor node comprises of Node MCU, Power supply, LCD, Relay board, soil sensor and Ultrasonic sensor. The system is designed to establish control and communication with specific agricultural field to take sensory data from soil and ultrasonic level sensor and control the PUMP IN motor and PUMP OUT motor with the help of mobile app. Fig.4.1 shows a block diagram for a system to control specific agricultural field with mobile app and IoT modem.

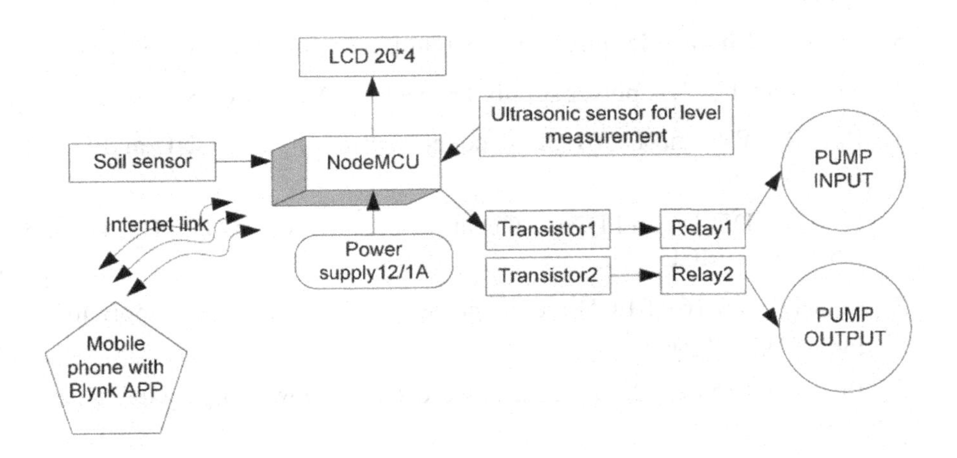

Fig.4.1: Block diagram of the system

Table:4.1 Components list

Component/Specification	Quantity
Power supply 12V/1Amp	1
2 Relay Board	1
Jumper wire M-M	20
Jumper wire M-F	20
Jumper wire F-F	20
Power supply extension (To get more +5V and GND)	1
LCD20*4	1
LCD patch/explorer board	1
NodeMCU patch	1
NodeMCU	1
Soil moisture sensor	1
Ultrasonic sensor patch	1

Note: All components are available at www.nuttyengineer.com

4.2 Circuit Diagram

Connection

1. Connect **SOIL sensor** output pin OUTPUT_SS to pinA0 of Arduino Nano
2. Connect +Vcc and GND pins of SOIL sensor to +5V and GND of Power supply
3. Connect **Ultrasonic sensor** RX-output pin to pinRX of NodeMCU
4. Connect +12V/1A power supply DC jack to DC jack of NodeMCU
5. Pins RS, RW and E of LCD is connected to pins D0, GND and D1 of NodeMCU.
6. Pins D4, D5, D6 and D7of LCD are connected to pins D2, D3, D4 and D5 of NodeMCU.
7. Pins 1,3 and 16 of LCD are connected to GND of power supply using power supply patch.
8. Pins 2 and 15 of LCD are connected to +5V of power supply using power supply patch.
9. Water Pump IN motor and Water Pump OUT motor to D6and D7 pins of

NodeMCU using relay board.

10. The base of NPN transistor 2N2222 is to be connected with pins of no-deMCU, in this case four pins D6,D7,D8,D9.
11. Emitter of transistor is grounded.
12. Collector of transistor is to be connected with L2 of relay and Li of relay

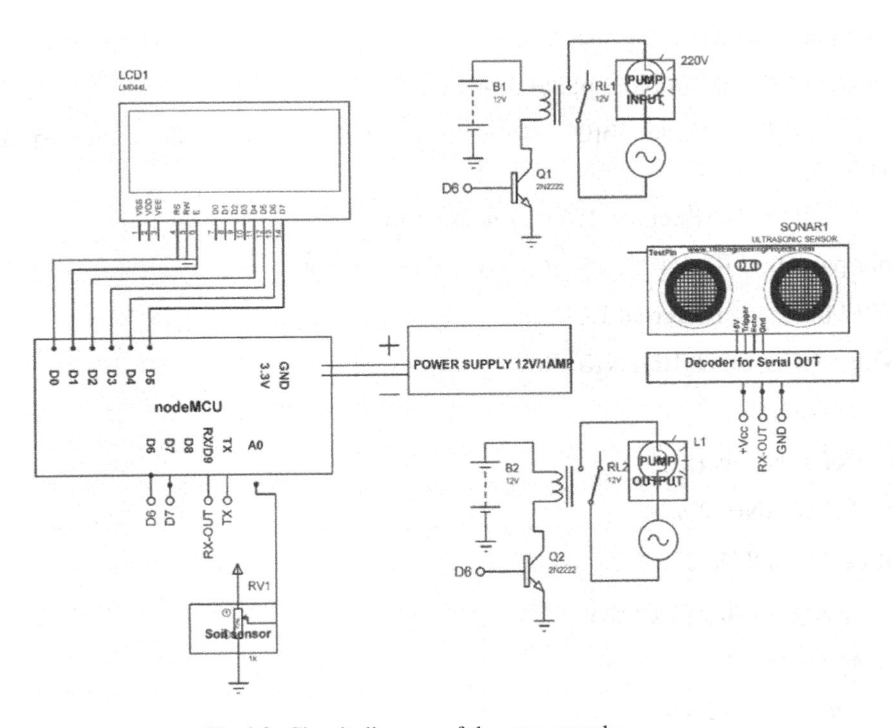

Fig.4.2: Circuit diagram of the sensor node

to positive terminal of 12V battery.

13. Negative terminal of battery is connected with ground.
14. One terminal of appliance (pump motor) is connected with 'NO' of relay and other to one end the AC source.
15. Other end of AC source is connected to 'Common' terminal of relay.

4.3 Program

```
#define BLYNK_PRINT Serial
///// library for external LCD
#include <LiquidCrystal.h>
LiquidCrystal lcd(D0, D1, D2, D3, D4, D5);
////// library for NodeMCU
#include <ESP8266WiFi.h>
#include <BlynkSimpleEsp8266.h>
char auth[] = "5c8e33bf09a04b03b2fa153928b075c5";///Taken recived at main
char ssid[] = "ESPServer_RAJ";// hotspot id
char pass[] = "RAJ@12345";// parsword of hotspot
//////// library for internal LCD
WidgetLCD LCD_BLYNK(V6);
///// for timer
BlynkTimer timer;
int PUMP_IN=12;//D6
int PUMP_OUT=13;//D7
String inputString_ULTRA = "";
String ULTRA;
////////////////// use button
BLYNK_WRITE(V1)
{
int PUMP_IN_VAL = param.asInt();
if(PUMP_IN_VAL==HIGH)
{
lcd.clear();
```

```
digitalWrite(PUMP_IN,HIGH);
digitalWrite(PUMP_OUT,LOW);
////// external LCD with NOdeMCU
lcd.setCursor(0,0);
lcd.print("PUMP_In Tigger");
//// LCD blynk
LCD_BLYNK.print(0,0,"PUMP_In Tigger");
delay(10);
}
}
BLYNK_WRITE(V2)
{
int PUMP_OUT_VAL = param.asInt();
if(PUMP_OUT_VAL==HIGH)
{
lcd.clear();
digitalWrite(PUMP_IN,LOW);
digitalWrite(PUMP_OUT,HIGH);
////// external LCD with nodeMCU
lcd.setCursor(0,0);
lcd.print("PUMP_OUT Tigger");
//// LCD blynk
LCD_BLYNK.print(0,0,"PUMP_OUT Tigger");
delay(10);
}
}
BLYNK_WRITE(V3)
{
```

```
int BOTH_ON = param.asInt();
if(BOTH_ON==HIGH)
{
lcd.clear();
digitalWrite(PUMP_IN,HIGH);
   digitalWrite(PUMP_OUT,HIGH);
   ////// external LCD with nodeMCU
   lcd.setCursor(0,0);
   lcd.print("BOTH ON");
    //// LCD blynk
   LCD_BLYNK.print(0,0,"BOTH ON");
   delay(10);
   }
 }
/////// read analog sensor
void READ_SENSOR()
{
 int READ_SOIL =analogRead(A0);
 ULTRASONIC_READ();
 Blynk.virtualWrite(V4,READ_SOIL);
 Blynk.virtualWrite(V5,ULTRA);
 lcd.setCursor(0,1);
 lcd.print("SOIL:");
 lcd.setCursor(5,1);
 lcd.print(READ_SOIL);
 lcd.setCursor(0,2);
 lcd.print("LEVEL:");
 lcd.setCursor(6,2);
```

```
lcd.print(ULTRA);
}
void setup()
{
Serial.begin(9600);
lcd.begin(20, 4);
Blynk.begin(auth, ssid, pass);
pinMode(PUMP_IN,OUTPUT);//D6 pin of NodeMCU
pinMode(PUMP_OUT,OUTPUT);//D7 pin of NodeMCU
timer.setInterval(1000L,READ_SENSOR);//// read sensor with setting delay
of 1 Sec
}
void loop()
{
Blynk.run();
timer.run(); // Initiates BlynkTimer
}
void ULTRASONIC_READ()
{
while (Serial.available()>0)
{
inputString_ULTRA = Serial.readStringUntil('\r');// Get serial input
ULTRA=String(((inputString_ULTRA[0]-48)*100) + ((inputString_ULTRA
[1]-
48)*10)+((inputString_ULTRA[2]-48)*1))+"."+String(((inputString_
ULTRA[4]-48)*10)+((inputString_ULTRA[5]-48)*1));
}
inputString_ULTRA = "";
delay(20);
}
```

4.4 Blynk APP

Blynk is iOS and Android platform to design mobile app. To design the app download latest Blynk library from:

Mobile App can easily be designed just by dragging and dropping widgets on the provided space.

Tutorials can be downloaded from: http://www.blynk.cc

Steps to design Blynk App

Step1: Download and install the Blynk App for your mobile Android or iphone from http://www.blynk.cc/getting-started/

Fig. 4.3: Create a new project

Step2: Create a Blynk Account

Step3: Create a new project

Click on + for creating new project and choose the theme dark (black background) or light (white background) and click on create.

Step4: Auth token is a unique identifier which will be received on the email address user provide at time of making account. Save this token, as this is required to copy in the main program of receiver section.

Step5: Select the device to which smart phone needs to communicate e.g.ESP8266 (NodeMCU)

Step6: Open widget box and select the components required for the project. For this project five buttons are selected.

Step7: Tap on the widget to get its settings, select virtual terminals as V1, V2 for each buttons, which needs to be defined later on the program.

Step8: After completing the widget settings, Run the project

Front end of APP for the proposed system

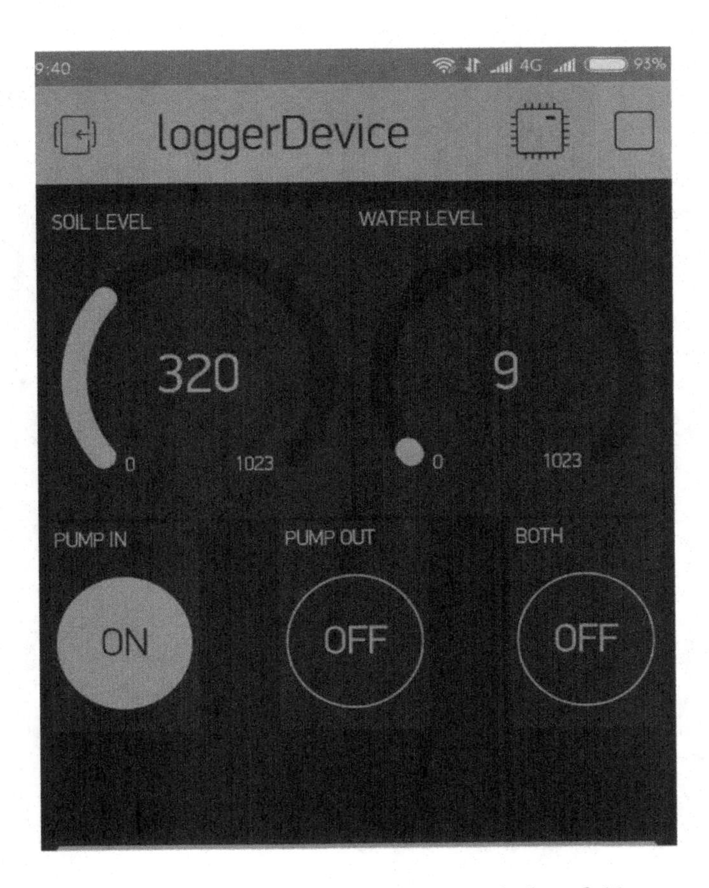

Fig. 4.4: Mobile App to automate agriculture field

5

Development of Local Web Server for Automation

5.1 Introduction

The chapter describes the development of local web server on the laptop, to control the actuator from the server. Web server is used as remote control for the system. The system is designed to control water pump with respect to the soil sensor values. It comprises of NodeMCU, soil sensor, LCD, two relay boards for pump in and out the water from field.

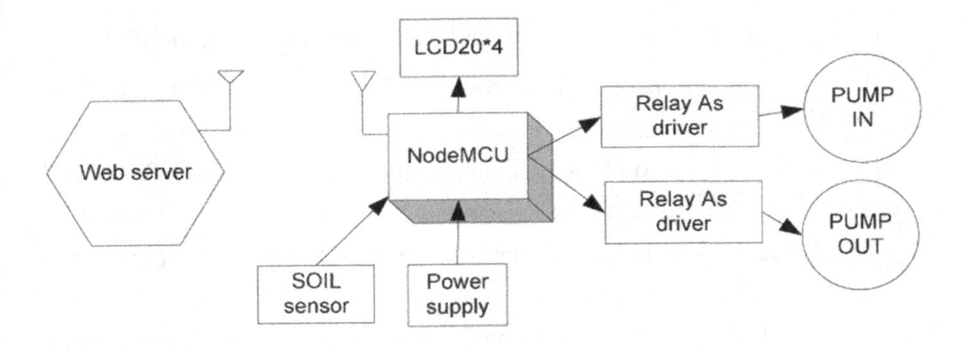

Fig. 5.1: Block diagram of the concept proof

Table 5.1: Components list

Component/Specification	Quantity
Power supply 12V/1Amp	1
2 Relay Board	1
Jumper wire M-M	20
Jumper wire M-F	20
Jumper wire F-F	20
Power supply extension (To get more +5V and GND)	1
LCD20*4	1
LCD patch/explorer board	1
NodeMCU patch	1
NodeMCU	1
Soil moisture sensor	1

Note: All components are available at www.nuttyengineer.com

5.2 Circuit Diagram

Connection

1. Connect SOIL sensor output pin OUT to pinA0 of NodeMCU
2. Connect +Vcc and GND pins of SOIL sensor to +5V and GND of Power supply
3. Connect +12V/1A power supply DC jack to DC jack of NodeMCU
4. Pins RS, RW and E of LCD is connected to pins D0, GND and D1 of NodeMCU.
5. Pins D4, D5, D6 and D7of LCD are connected to pins D2, D3, D4 and D5 of NodeMCU.
6. Pins 1,3 and 16 of LCD are connected to GND of power supply using power supply patch.
7. Pins 2 and 15 of LCD are connected to +5V of power supply using power supply patch.
8. Water Pump IN motor and Water Pump OUT motor to D6and D7 pins of NodeMCU using relay board.
9. The base of NPN transistor 2N2222 is to be connected with pins of nodeMCU, in this case four pins D6,D7,D8,D9.
10. Emitter of transistor is grounded.

11. Collector of transistor is to be connected with L2 of relay and Li of relay to positive terminal of 12V battery.
12. Negative terminal of battery is connected with ground.
13. One terminal of appliance (pump motor) is connected with 'NO' of relay and other to one end the AC source.
14. Other end of AC source is connected to 'Common' terminal of relay.

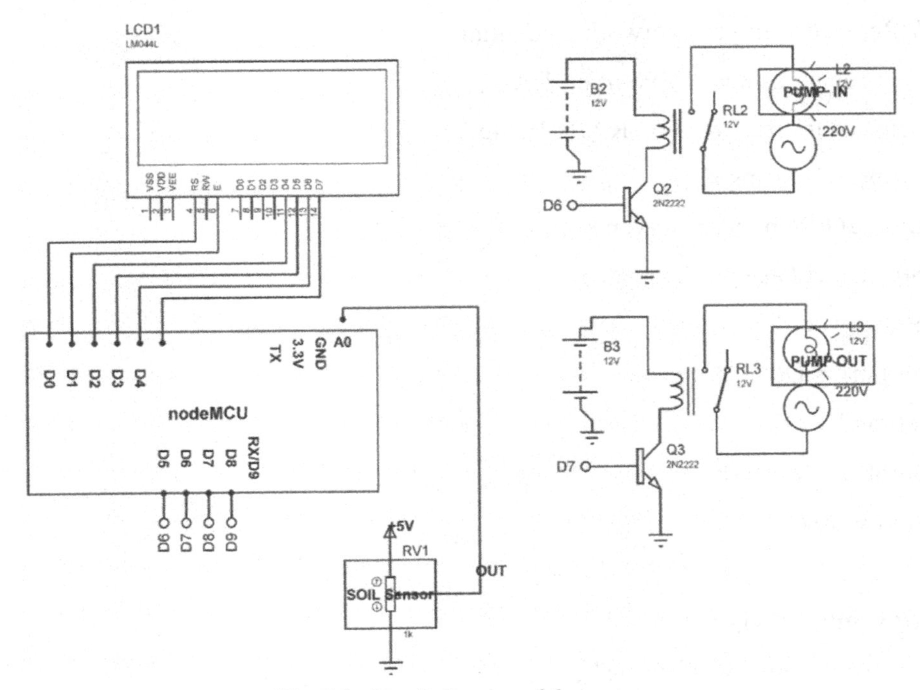

Fig. 5.2: Circuit diagram of the system

5.3 Program

Code for developing local webserver

```
////////////for hot spot

#include <ESP8266WiFi.h>

#include <WiFiClient.h>

#include <ESP8266WebServer.h>

#include <ESP8266mDNS.h>
```

```
int Analog = A0;
#include <LiquidCrystal.h>
// initialize the library with the numbers of the interface pins
LiquidCrystal lcd(D0, D1, D2, D3, D4, D5);
////////////for hotspot
MDNSResponder mdns;
// Replace with your network credentials
const char* ssid = "ESPServer_RAJ";
const char* password = "RAJ@12345";
String webString="";
ESP8266WebServer server(80);
String webPage = "";
String web="";
int pin1 = D6;
int pin2 = D7;
int SOIL_level=0;
void setup()
{
lcd.begin(20, 4);
lcd.print("robot Monitoring");
webPage +="<h2>ESP8266 Web Server new</h2><p>SOIL METER <a
href=\"SOIL\"><button> SOIL LEVEL</button></a></p>";// for soil sensor
webPage += "<p>PUMP IN-STATUS <a
href=\"PUMPINON\"><button>ON</button></a> <a
href=\"PUMPINOFF\"><button>OFF</button></a></p>";
webPage += "<p>PUMP OUT-STATUS <a
href=\"PUMPOUTON\"><button>ON</button></a> <a
href=\"PUMPOUTOFF\"><button>OFF</button></a></p>";
// preparing GPIOs
pinMode(pin1, OUTPUT);
```

```
digitalWrite(pin1, LOW);
pinMode(pin2, OUTPUT);
digitalWrite(pin2, LOW);
delay(1000);
Serial.begin(115200);
WiFi.begin(ssid, password);
Serial.println("");
// Wait for connection
while (WiFi.status() != WL_CONNECTED)
{
delay(500);
Serial.print(".");
}
Serial.println("");
Serial.print("Connected to ");
Serial.println(ssid);
Serial.print("IP address: ");
Serial.println(WiFi.localIP());
if (mdns.begin("esp8266", WiFi.localIP()))
{
Serial.println("MDNS responder started");
}
server.on("/", []()
{
server.send(200, "text/html", webPage);
});
/*****************************************************************
*************************/
```

```
server.on("/SOIL", []()
{
get_SOIL();
webString="SOIL: "+String((float)SOIL_level)+"oC";
server.send(200, "text/plain", webString); // send to someones browser when
asked
});
server.on("/PUMPINON", []()
{
server.send(200, "text/html", webPage);
digitalWrite(pin1, HIGH);
digitalWrite(pin2, LOW);
lcd.clear();
lcd.setCursor(0, 1);
lcd.print("PUMP IN ON ");
delay(1000);
});
server.on("/PUMPINOFF", []()
{
server.send(200, "text/html", webPage);
digitalWrite(pin1, LOW);
digitalWrite(pin2, LOW);
lcd.clear();
lcd.setCursor(0, 1);
lcd.print("PUMP IN OFF");
delay(1000);
});
server.on("/PUMPOUTON", []()
{
```

```
server.send(200, "text/html", webPage);
digitalWrite(pin1, LOW);
digitalWrite(pin2, HIGH);
lcd.clear();
lcd.setCursor(0, 1);
lcd.print("PUMP OUT ON ");
delay(1000);
});
server.on("/PUMPOUTOFF", []()
{
server.send(200, "text/html", webPage);
digitalWrite(pin1, LOW);
digitalWrite(pin2, LOW);
lcd.clear();
lcd.setCursor(0, 1);
lcd.print("PUMP OUT OFF ");
delay(1000);
});
server.begin();
Serial.println("Congats for connection, Your HTTP server started");
}
void loop()
{
server.handleClient();
get_SOIL();
lcd.clear();
lcd.setCursor(0, 0);
lcd.print(SOIL_level);
```

```
delay(500);
}
void get_SOIL()
{
int SOIL_level1= analogRead(Analog);
SOIL_level=SOIL_level1/2;
}
```

5.4 Local Web server

After running the program on Arduino IDE

Check serial monitor of Arduino and check IP address of device on it.

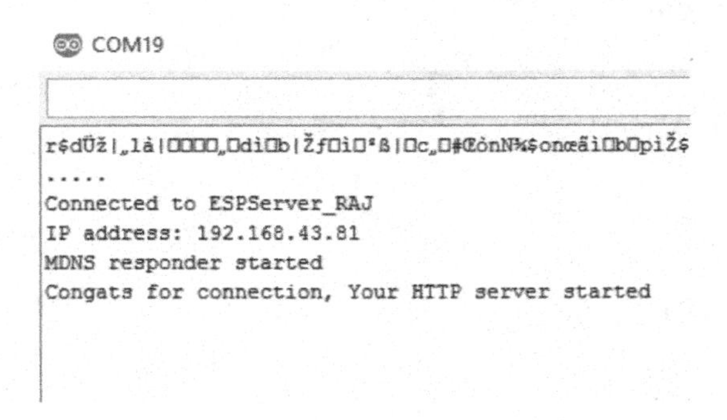

Fig. 5.3: Serial COM Port

When same IP address is typed on web browser, it will give the local webserver for the system. Fig.5.4 shows the snapshot for the local web server and device control window available at 191.168.43.81.

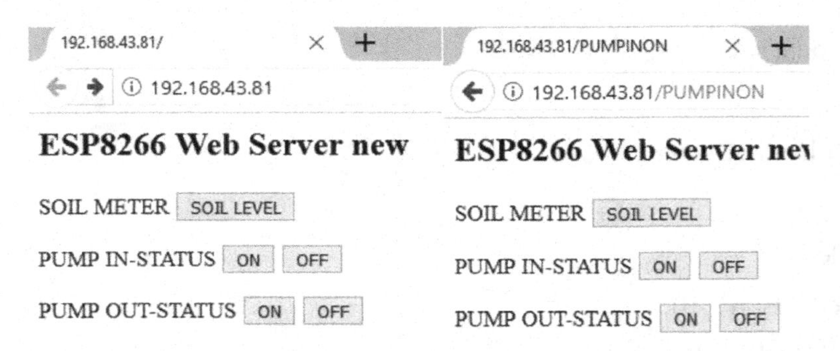

Fig. 5.4: Snapshot of the local webserver

6

LabVIEW Based Data Logger for Agricultural Field Parameters Monitoring System

6.1 Introduction

This chapter describes the development of data logger of agricultural field parameter monitoring system with LabVIEW. The system comprises of two sections- Fields device and data logger. Field device comprises of Arduino, smoke detector (for sensing hazardous gases), soil moisture sensor, temperature sensor, humidity sensor, water level sensor, battery, display unit, 2.4 GHz RF modem. Data logger comprises of Arduino, display unit, 2.4 GHz RF modem, power supply, PC with LabVIEW. The system is designed to monitor the environment and agriculture field parameters. The field device is designed to collect all physical parameters and communicate it to the receiver end with LabVIEW. For establishing communication between two sections 2.4GHz RF modem is used, which can be replaced with any other modem also. Fig.6.1 shows block diagram of field device and Fig.6.2 shows block diagram of data logger. Table.6.1 & Table.6.2 shows the component list required to develop the field device and data logger.

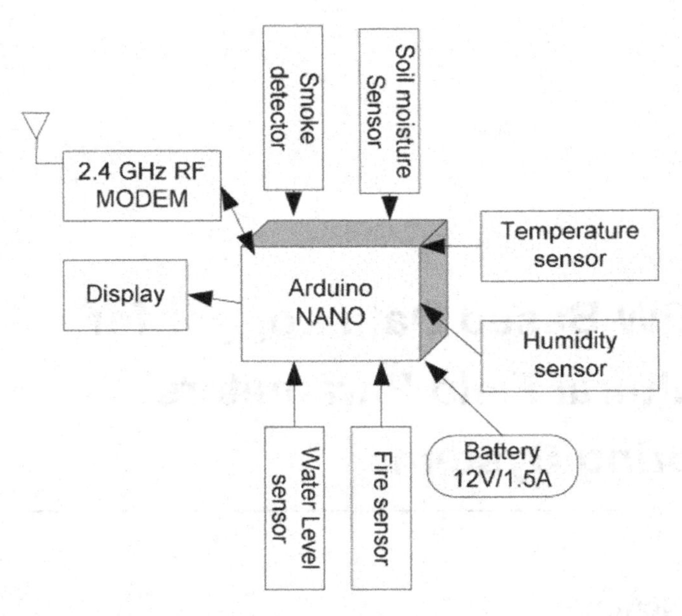

Fig.6.1: Block diagram of field device

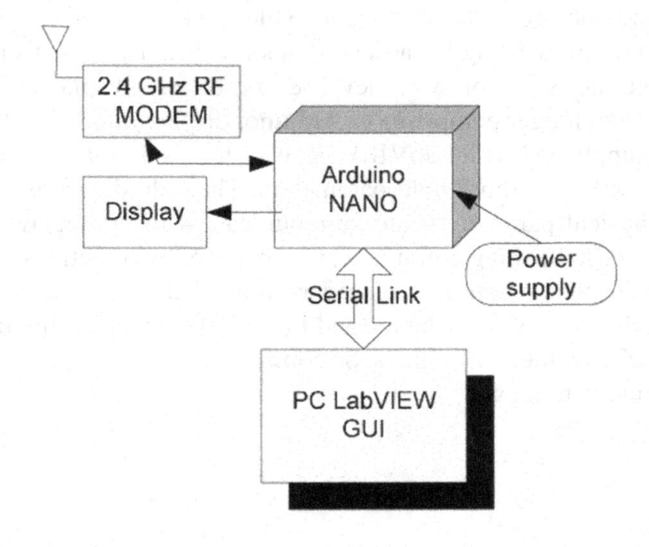

Fig. 6.2: Data logger with LabVIEW

Table 6.1: Components list for field device

Component	Quantity
Power supply 12V/1.5Amp	1
Arduino NANO	1
RF 2.4GHz modem	1
Jumper wire M-M	20
Jumper wire M-F	20
Jumper wire F-F	20
Power supply extension (To get more +5V and GND)	1
Water level sensor	1
Fire sensor	1
Soil sensor	1
Light sensor	1
DHT11	1
RF modem explorer board	1
Smoke detector	1
LCD 20*4	1
LCD patch	1

Table:6.2 Components list for data logger

Component/Specification	Quantity
Power supply 12V/1Amp	1
Arduino Nano	1
RF 2.4GHz modem	1
Jumper wire M-M	20
Jumper wire M-F	20
Jumper wire F-F	20
Power supply extension (To get more +5V and GND)	1
LCD20*4	1
LCD patch/explorer board	1
5 Push button array	1
RF modem explorer board	1

Note: All components are available at www.nuttyengineer.com

6.2 Circuit Diagram

6.2.1 Field device connections

1. Connect **Fire sensor** output pin OUTPUT_FS to pin13 of Arduino Nano
2. Connect +Vcc and GND pins of sensors to +5V and GND of power supply
3. Connect **Smoke detector sensor** output pin OUTPUT_RS to pinA0 of Arduino Nano
4. Connect +Vcc and GND pins of smoke detector sensors to +5V and GND of Power supply
5. Connect **SOIL sensor** output pin OUTPUT_SS to pinA1 of Arduino Nano
6. Connect +Vcc and GND pins of SOIL sensor to +5V and GND of Power supply
7. Connect **Light intensity sensor** output pin OUTPUT_LS to pinA2 of Arduino Nano
8. Connect +Vcc and GND pins of Light intensity sensor to +5V and GND of Power supply
9. Connect pin 2 of **DHT11** to pin 2 of Arduino Nano
10. Connect +Vcc and GND pins of DHT11 sensor to +5V and GND of Power supply
11. Connect RX out or data out pin of **water level sensor** to RX(0) pin of Arduino Nano
12. Connect +Vcc and GND pins of water level sensor to +5V and GND of Power supply
13. Connect TX, RX, +Vcc and GND pins of RF modem to pins 6, 7, +5V and GND of Arduino Nano.
14. Connect +12V/1.5A battery DC jack to DC jack of Arduino Nano.

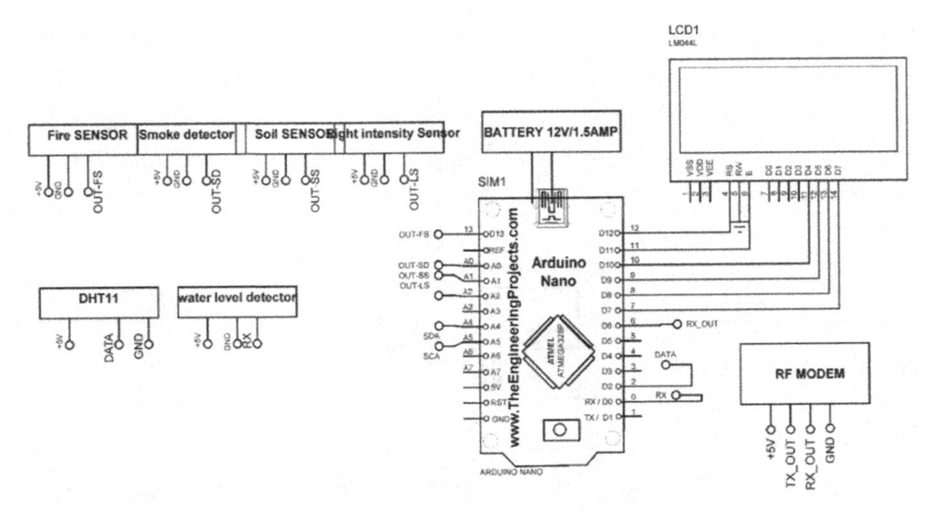

Fig. 6.3: Circuit diagram of field device

6.2.2 LabVIEW based Data logger connection

1. Connect TX, RX, +Vcc and GND pins of RF modem to pins 6, 7, +5V and GND of Arduino Nano.
2. Connect +12V/1A power supply DC jack to DC jack of Arduino Nano.
3. Connect D7 and D8 pins of NodeMCU to TX and RX pins of Arduino Nano
4. Pins 1,3 and 16 of LCD are connected to GND of power supply using power supply patch.
5. Pins 2 and 15 of LCD are connected to +5V of power supply using power supply patch.

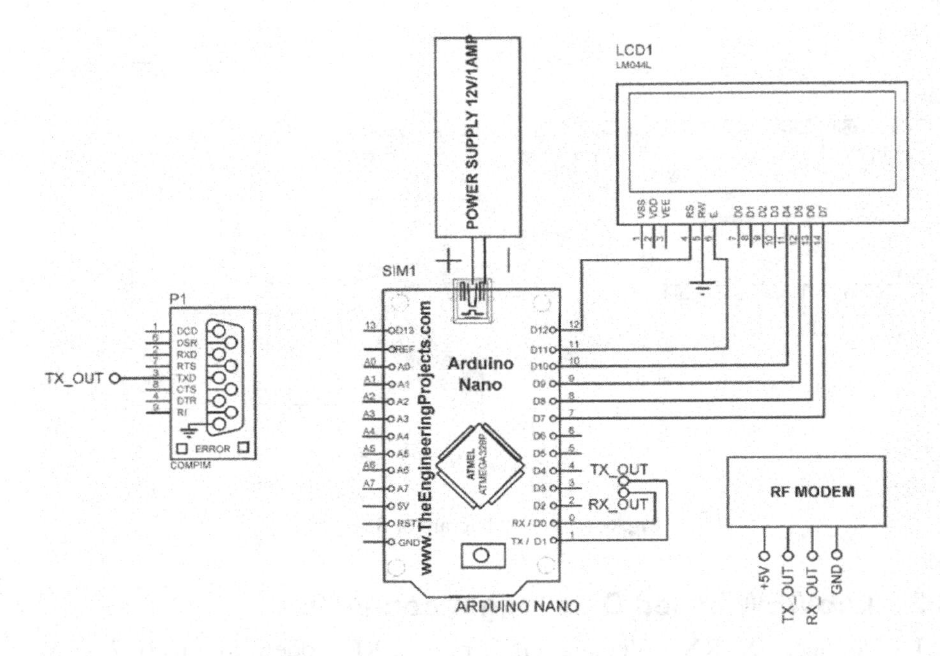

Fig.6.4: Circuit diagram of LabVIEW based data logger

6.3 Program

6.3.1 Field device code

```
#include <dht.h>
dht DHT;
#define DHT11_PIN 2
/////////////// library for LCD
#include <LiquidCrystal.h>
LiquidCrystal lcd(13, 12, 11, 10, 9, 8);
////////////// libray for Softserial
#include <SoftwareSerial.h>
SoftwareSerial mySerial(5,6);// 5 rx /6 tx
 int Fire_level,SOIL_level,LIGHT_level,SMOKE_level;
String inputString_ULTRA = "";
String ULTRA;
```

```
void setup()
{
Serial.begin(9600);
mySerial.begin(9600);
lcd.begin(20, 4);
}
 void loop()
{
lcd.clear();
//////// read Fire sensor
Fire_level=digitalRead(13);
//////// read Soil sensor
SOIL_level=analogRead(A0);
SOIL_level=SOIL_level/2;
//////// read light sensor
LIGHT_level=analogRead(A1);
//////// read RAIN sensor
SMOKE_level=analogRead(A2);
//// read DHT sensor
int chk = DHT.read11(DHT11_PIN);
ULTRASONIC_READ();
 if(Fire_level==LOW)
{
int FIRE_level=10;
/////////// soil sensor
lcd.setCursor(0,0);
lcd.print("SOIL:");
lcd.print(SOIL_level);
```

```
////// read air quality sensor
lcd.setCursor(10,0);
lcd.print("LIGT:");
lcd.print(LIGHT_level);
//////// read rain sensor level
lcd.setCursor(0,1);
lcd.print("RAIN:");
lcd.print(SMOKE_level);
/////// fire
lcd.setCursor(10,1);
lcd.print("FStatus:");
lcd.print("Y");
////// read and Display DHT
lcd.setCursor(0,2);
lcd.print("T:");
lcd.print(DHT.temperature);
lcd.setCursor(10,2);
lcd.print("H:");
lcd.print(DHT.humidity);
lcd.setCursor(0,3);
lcd.print("WATER_LEVEL:");
lcd.print(ULTRA);
Serial.print(SOIL_level);
Serial.print(",");
Serial.print(LIGHT_level);
Serial.print(",");
Serial.print(FIRE_level);
Serial.print(",");
```

```
Serial.print(SMOKE_level);
Serial.print(",");
Serial.print(DHT.temperature);
Serial.print(",");
Serial.print(DHT.humidity);
Serial.print(",");
Serial.print(ULTRA);
Serial.print('\n');
delay(30);
mySerial.print(SOIL_level);
mySerial.print(",");
mySerial.print(LIGHT_level);
mySerial.print(",");
mySerial.print(FIRE_level);
mySerial.print(",");
mySerial.print(SMOKE_level);
mySerial.print(",");
mySerial.print(DHT.temperature);
mySerial.print(",");
mySerial.print(DHT.humidity);
mySerial.print(",");
mySerial.print(ULTRA);
mySerial.print('\n');
delay(30);
}
else
{
 int FIRE_level=20;
```

```
/////////// soil sensor
/////////// soil sensor
lcd.setCursor(0,0);
lcd.print("SOIL:");
lcd.print(SOIL_level);
////// read air quality sensor
lcd.setCursor(10,0);
lcd.print("LIGT:");
lcd.print(LIGHT_level);
//////// read rain sensor level
lcd.setCursor(0,1);
lcd.print("SMOKE:");
lcd.print(SMOKE_level);
/////// fire
lcd.setCursor(10,1);
lcd.print("FStatus:");
lcd.print("Y");
////// read and Display DHT
lcd.setCursor(0,2);
lcd.print("T:");
lcd.print(DHT.temperature);
lcd.setCursor(10,2);
lcd.print("H:");
lcd.print(DHT.humidity);
lcd.setCursor(0,3);
lcd.print("WATER_LEVEL:");
lcd.print(ULTRA);
Serial.print(SOIL_level);
```

```
Serial.print(",");
Serial.print(LIGHT_level);
Serial.print(",");
Serial.print(FIRE_level);
Serial.print(",");
Serial.print(SMOKE_level);
Serial.print(",");
Serial.print(DHT.temperature);
Serial.print(",");
Serial.print(DHT.humidity);
Serial.print(",");
Serial.print(ULTRA);
Serial.print('\n');
delay(30);
mySerial.print(SOIL_level);
mySerial.print(",");
mySerial.print(LIGHT_level);
mySerial.print(",");
mySerial.print(FIRE_level);
mySerial.print(",");
mySerial.print(SMOKE_level);
mySerial.print(",");
mySerial.print(DHT.temperature);
mySerial.print(",");
mySerial.print(DHT.humidity);
mySerial.print(",");
mySerial.print(ULTRA);
mySerial.print('\n');
```

```
delay(30);
}
}
void ULTRASONIC_READ()
{
while (mySerial.available()>0)
{
inputString_ULTRA = mySerial.readStringUntil('\r');// Get serial input
ULTRA=String(((inputString_ULTRA[0]-48)*100) + ((inputString_ULTRA[1]-48)*10)+((inputString_ULTRA[2]-48)*1))+"."+String(((inputString_ULTRA[4]-48)*10)+((inputString_ULTRA[5]-48)*1));
}
inputString_ULTRA = "";
delay(20);
}
```

6.3.2 Data logger Code

```
#include "StringSplitter.h"
/////////////// library for LCD
#include <LiquidCrystal.h>
LiquidCrystal lcd(12, 11, 10, 9, 8,7);
String TEMP_HUM_STRING = "";         // a string to hold incoming data
String SOIL_level,LIGHT_level,SMOKE_level,FIRE_level,TEMP_level,HUM_level,PRESS_level,ALT_level;
void setup()
{
Serial.begin(9600);
lcd.begin(20, 4);
}
void loop()
```

```
{
lcd.setCursor(0,0);
lcd.print("SOIL:");
lcd.print(SOIL_level);
////// read air quality sensor
lcd.setCursor(10,0);
lcd.print("LIGT:");
lcd.print(LIGHT_level);
//////// read rain sensor level
lcd.setCursor(0,1);
lcd.print("SMOKE:");
lcd.print(SMOKE_level);
/////// fire
lcd.setCursor(10,1);
lcd.print("FStatus:");
lcd.print("Y");
////// read and Display DHT
lcd.setCursor(0,2);
lcd.print("T:");
lcd.print(TEMP_level);
lcd.setCursor(10,2);
lcd.print("H:");
lcd.print(HUM_level);
lcd.setCursor(0,3);
lcd.print("WATER_LEVEL:");
lcd.print(WATER_LEVEL);
////// serial data to LabVIEW
Serial.print("SOIL:");
```

```
Serial.print(SOIL_level);
Serial.print("LIGHT:");
Serial.print(LIGHT_level);
Serial.print("FIRE:");
Serial.print(FIRE_level);
Serial.print("SMOKE:");
Serial.print(SMOKE_level);
Serial.print("TEMP:");
Serial.print(TEMP_level);
Serial.print("HUM:");
Serial.print(HUM_level);
Serial.print("WATER:");
Serial.println(WATER_level);
}
void serialEvent_NODEMCU()
{
while (Serial.available()>0)
{
TEMP_HUM_STRING = Serial.readStringUntil('\n');// Get serial input
StringSplitter *splitter = new StringSplitter(TEMP_HUM_STRING, ',', 7);
int itemCount = splitter->getItemCount();
for(int i = 0; i < itemCount; i++)
{
String item = splitter->getItemAtIndex(i);
SOIL_level = splitter->getItemAtIndex(0);
LIGHT_level = splitter->getItemAtIndex(1);
FIRE_level = splitter->getItemAtIndex(2);
SMOKE_level= splitter->getItemAtIndex(3);
```

```
TEMP_level=splitter->getItemAtIndex(4);
HUM_level=splitter->getItemAtIndex(5);
WATER_level=splitter->getItemAtIndex(6);
}
TEMP_HUM_STRING= "";
delay(20);
}
}
```

6.4 LabVIEW GUI

6.4.1 Steps to design LabVIEW GUI

Step1: Install the LabVIEW software

The first step is to install LabVIEW software. Download and install the NI VISA driver, It VISA is required for serial communication.

Step 2: Create 'New Project'

Open the LabVIEW window and click on the 'Create Project' button.

Select Blank VI from the list of items and click Finish. A blank front panel window and block diagram window will appear.

Step 3: Running and Debugging VIs

To run a VI, connect all the sub VIs, functions, and structures as per the data types for the terminals.

6.4.2 Design steps for front Panel

The front panel is the user interface of a VI. Front panel is designed by right click on blank VI and select the controls and indicators. Select controls and indicators from the Controls palette and place them on the front panel. Controls are knobs, push buttons, dials, and other input mechanisms. Indicators are graphs, LEDs, and other output displays.

6.4.3 Building the Block Diagram

Block diagram is a code using graphical representations of functions to control the front panel objects. The block diagram contains the graphical source code, also known as 'G code'. The components can be selected by right click on front panel or go to View and select 'Function Pallete'.

6.4.4 Virtual Instrument Software Architecture (VISA)

The Virtual Instrument Software Architecture (VISA) is a standard for configuring, programming, and troubleshooting instrumentation systems. VISA offers the programming interface between the hardware and development environments like LabVIEW. NI-VISA is the National Instruments implementation of the VISA I/O standard. NI-VISA includes software libraries.

Components used to design LabVIEW GUI

Visa Configure Serial port- This port initialize the serial port specified by VISA resource name with required settings.

VISA resource name (COM number)

Right click on it and select ***create*** then select ***control*** to choose appropriate COM port.

Baud rate-choose 9600

Data bits-choose 8 bits

Parity-choose none

Stop Bit-choose 1

Flow control – choose none

Visa resource name out- Connect this pin to **Visa resource name of VISA serial read block.**

Error out-connect this pin to error pin of **VISA serial read**

VISA serial Read

It reads the identified number of bytes from the device or interface identified by **VISA resource name** and sends the data in **read buffer**. It reads the data available at serial port from the device linked.

Byte Count- Right click on it and select ***create*** to the **indicator** to count the byte at serial port.

Read Buffer - Right click on it and select ***create*** to the ***indicator*** to check the string value at serial port.

Visa resource name out-Connect this pin to Visa resource name of VISA closeblock

Error out- Connect this pin to error pin of **VISA close.**

Error in- connect this pin to error out pin of VISA configure serial port.

Match Pattern

It searches for expression in string beginning at offset, and when it get the string it matches the string with predefined data.

String-Connect this pin to read buffer pin of **VISA read.**

Regular expression- Right click on it and select ***create*** to the ***constant***.

After substring-connect this pin to **string** of **Decimal String to Number** block and **create** to **constant.**

Decimal String to Number- Converts the numeric characters in **string,** starting at **offset**, to a decimal integer and returns it in **number.**

Number- Connect this pin to the input of waveform chart.

VISA Close - Closes a device session or event object specified by VISA resource name

VISA Serial Write- Writes the data from write buffer to the device or interface identified by VISA resource name

Visa Configure Serial port- It sets the serial port identified by VISA resource name to the specified settings.

VISA resource name (COM number)- Right click on it select ***create*** and then select **control** to choose appropriate COM port

Baud rate-choose 9600

Data bits-choose 8 bits

Parity-choose none

Stop Bit-choose 1

Flow control – choose none

Visa resource name out- Connect this pin to **Visa resource name of VISA serial read** block.

Error out-connect this pin to error pin of **VISA serial read**

VISA Serial Write- It writes the data from write buffer to the device or interface stated by VISA resource name.

VISA resource name- Connect this pin to **VISA resource name out** pin of

VISA configure serial port.

Write Buffer - It comprises of the data to be written to the device.Right click on it and select **create** to the **constant** to send string at serial port.

Visa resource name out- Connect this pin to **Visa resource name** of **VISA close** block

Error out- Connect this pin to error pin of **VISA close.**

Error in- connect this pin to error out pin of **VISA configure serial port.**

VISA Close - Closes a device session or event object specified by VISA resource name.

6.5 LabVIEW GUI for the system

By following the steps described in the 6.4 section, front panel and block diagram are designed, Fig.6.5 & 6.6.

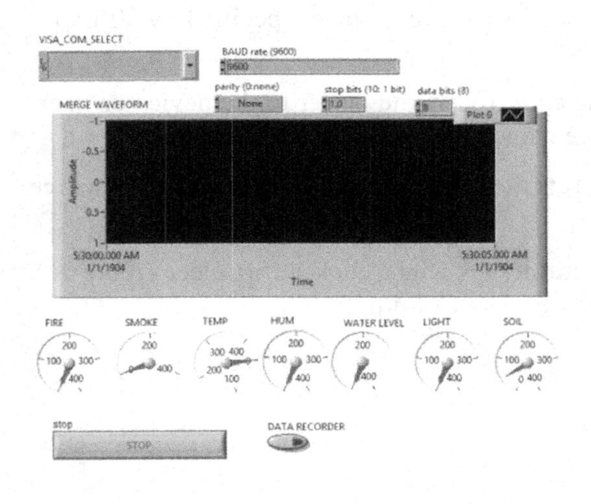

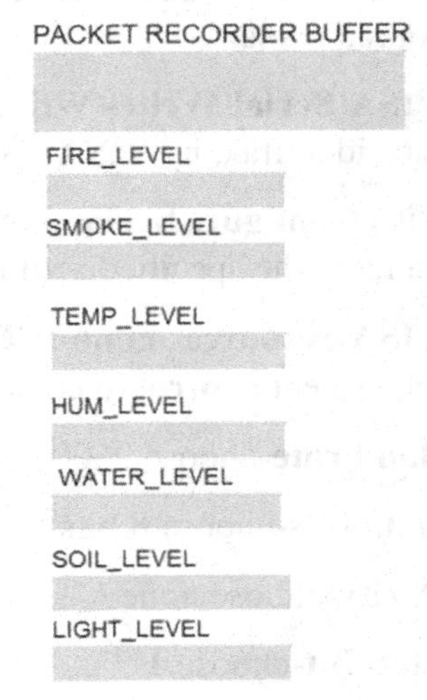

Fig. 6.5: LabVIEW front panel

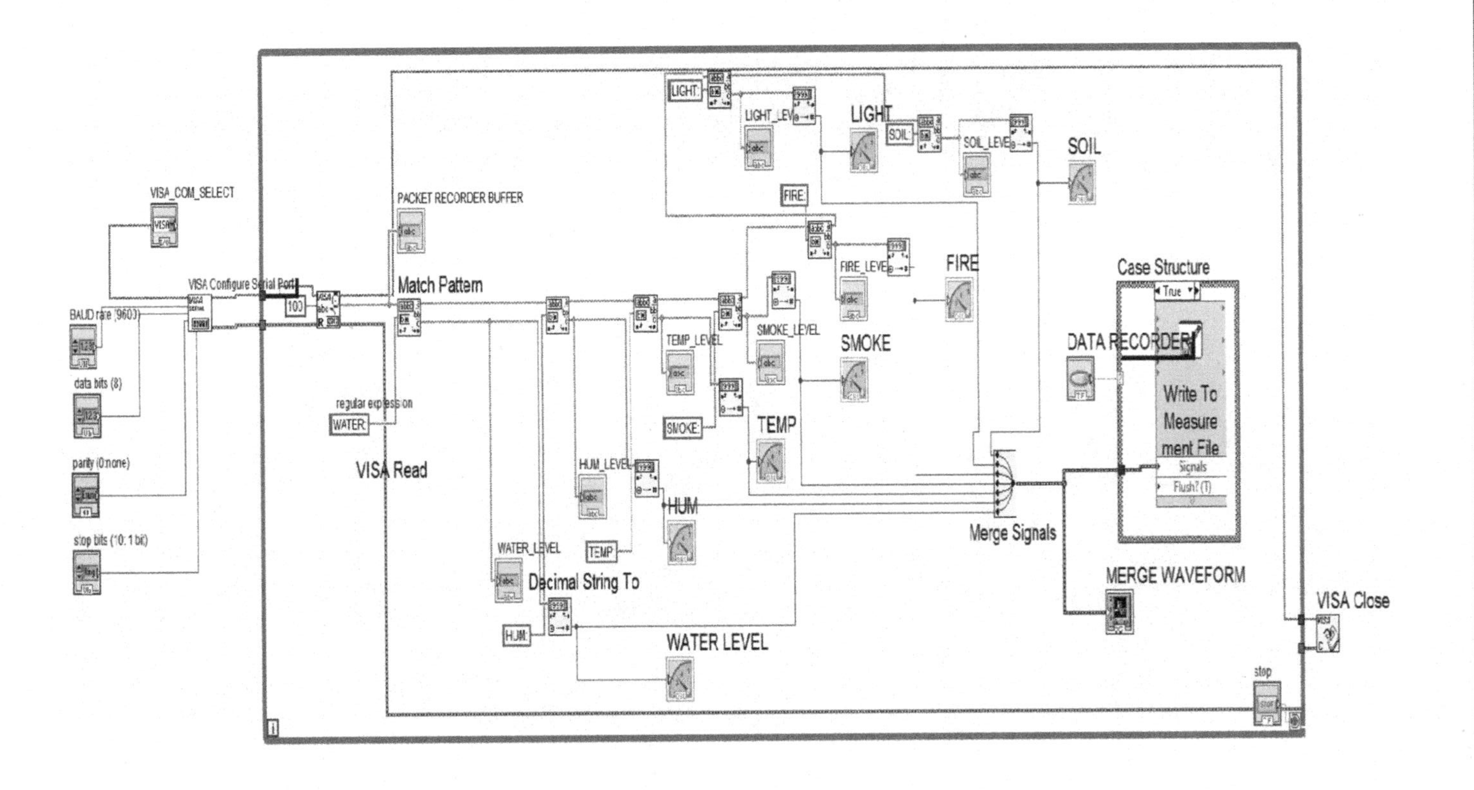

Fig. 6.6: LabVIEW front panel

7

MATLAB Based Data Logger for Agricultural Field Parameters Monitoring System

7.1 Introduction

This chapter describes the development of data logger of agricultural field parameter monitoring system with MATLAB. The system comprises of Arduino, water level sensor, humidity sensor, light sensor, soil moisture sensor, temperature sensor, smoke detector, power supply, PC with MATLAB. The system is designed to monitor the environment and agriculture field parameters. The field device is designed to collect all physical parameters and same is communicated serially to MATLAB with I/O package. Fig.7.1 shows block diagram of the system. Table.7.1 shows the component list required to develop the system.

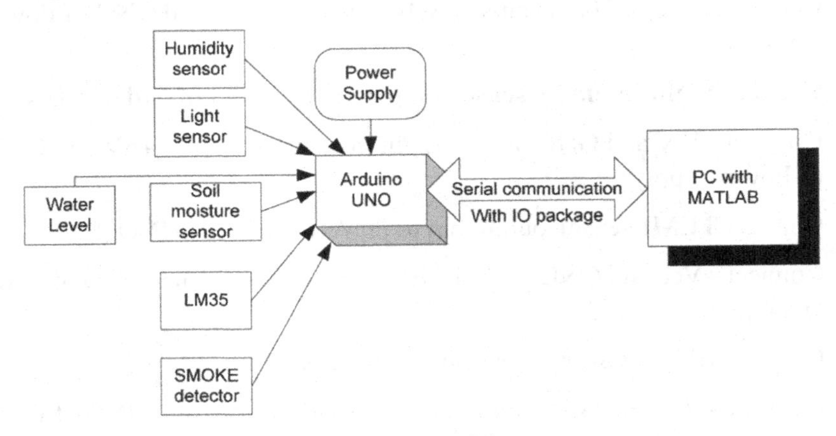

Fig. 7.1: Block diagram of the system

Table 7.1: Components list

Component	Quantity
Power supply 12V/1.5Amp	1
Arduino UNO	1
Jumper wire M-M	20
Jumper wire M-F	20
Jumper wire F-F	20
Power supply extension (To get more +5V and GND)	1
Water level sensor	1
Humidity sensor	1
Soil sensor	1
Light sensor	1
Temperature sensor	1
Smoke detector	1

Note: All components are available at www.nuttyengineer.com

7.2 Circuit Diagram

Connections

1. Connect **Smoke detector sensor** output pin to pinA0 of Arduino UNO
2. Connect +Vcc and GND pins of smoke detector sensors to +5V and GND of Power supply
3. Connect **SOIL sensor** output pin to pinA2of Arduino UNO
4. Connect +Vcc and GND pins of SOIL sensor to +5V and GND of Power supply
5. Connect **Light intensity sensor** output pin to pinA3 of Arduino UNO
6. Connect +Vcc and GND pins of Light intensity sensor to +5V and GND of Power supply
7. Connect **TEMPsensor** output pin to pinA1of Arduino UNO
8. Connect +Vcc and GND pins of TEMP sensor to +5V and GND of Power supply
9. Connect **HUM sensor** output pin to pinA5of Arduino UNO
10. Connect +Vcc and GND pins of HUM sensor to +5V and GND of Power supply

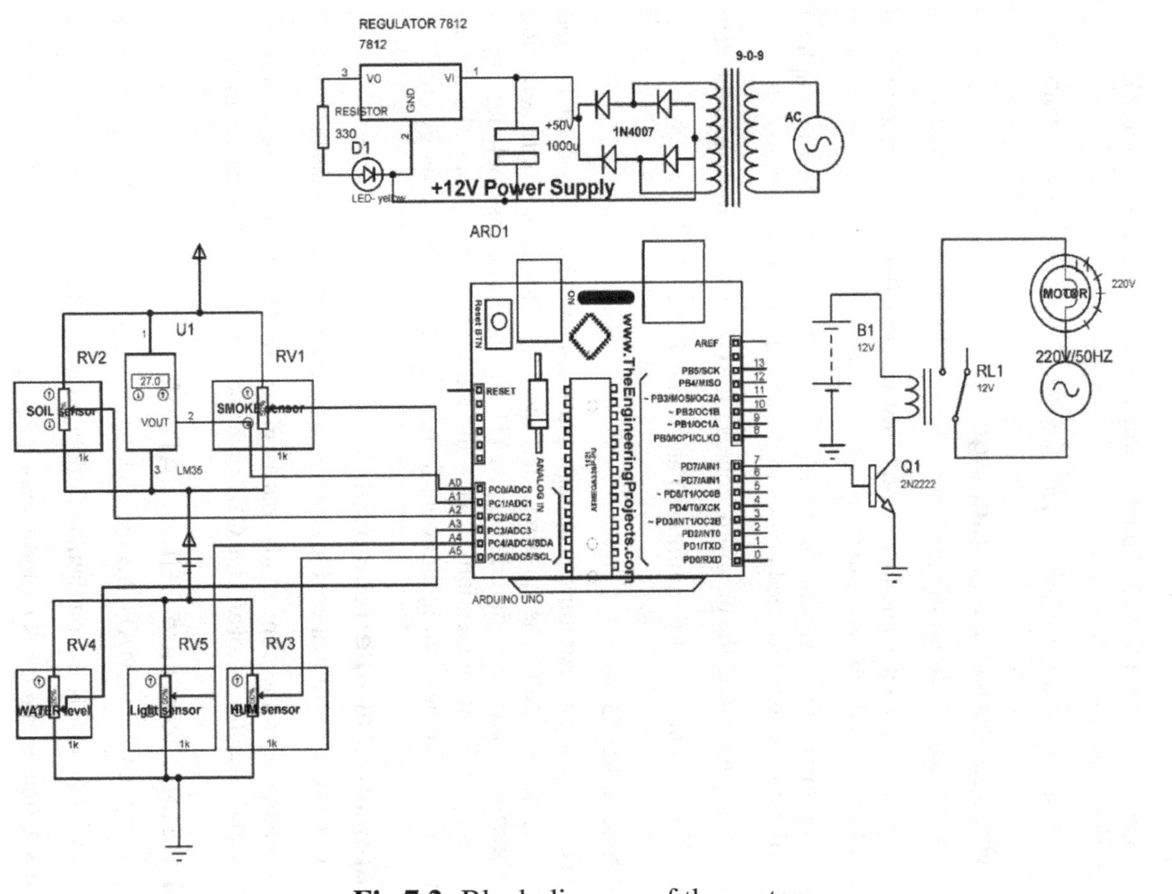

Fig.7.2: Block diagram of the system

11. Connect out pin of **water level sensor** to A4 pin of Arduino UNO
12. Connect +Vcc and GND pins of water level sensor to +5V and GND of Power supply
13. Connect +12V/1.5A battery DC jack to DC jack of Arduino UNO.

7.3 Procedure to run I/O package

1. Download Arduino I/O package
2. Open the package and install the firmware from the file install arduino.m from the matlab platform
3. Again open the Arduino I/O folder and open the pde folder and upload the pde program named adioes in the arduino microcontroller
4. After uploading the program, connect the GUI to matlab to read the data
5. The digital output is configured at the pin 7 of the Arduino.

7.4 MATLAB GUI

MATLAB is a programming language platform by MathWorks. It is used for, matrix manipulations, plotting of functions and data, implementation of algorithms and user interfaces. MATLAB applications include, signal processing and communications, Image and video processing, control systems, test and measurement, computational finance, computational biology etc.

Graphical user interface (GUI)

It is a MATLAB tool that enables a user to perform interactive tasks.

GUI manipulates the commands that is given by the end user and responds accordingly. Each control and the GUI have one or more *callbacks* as command.

7.4.1 Steps to create GUI in MATLAB

Step 1: Open GUI in MATLAB by clicking on icon.

Step 2: Click on OK button then a window will be opened.

Step 3: Click on the icons to add component on window.

Step 4: Double click on component a window will be opened, where some options like- font size, color and name can be assigned, then save it and repeat it for another components.

Step 7: Write the Function inside window shown in Fig.21.6 to open the COM port and data command.

Step 8: Write MATLAB code in the program window and check it on MATLAB GUI.

7.5 MATLAB Code

```
function varargout = DATAloggerMATLAB(varargin)
gui_Singleton = 1;
gui_State = struct('gui_Name',      mfilename, ...
'gui_Singleton',  gui_Singleton, ...
'gui_OpeningFcn', @DATAloggerMATLAB_OpeningFcn, ...
'gui_OutputFcn',  @DATAloggerMATLAB_OutputFcn, ...
'gui_LayoutFcn',  [] , ...
'gui_Callback',   []);
if nargin && ischar(varargin{1})
gui_State.gui_Callback = str2func(varargin{1});
end
if nargout
[varargout{1:nargout}] = gui_mainfcn(gui_State, varargin{:});
else
gui_mainfcn(gui_State, varargin{:});
end
% End initialization code - DO NOT EDIT
% --- Executes just before DATAloggerMATLAB is made visible.
function DATAloggerMATLAB_OpeningFcn(hObject, eventdata, handles,
varargin)
handles.output = hObject;
% Update handles structure
guidata(hObject, handles);
delete(instrfind({'Port'},{'COM38'}))
clear a;
global a;
```

```
global stop;
stop='e';
global entry;
entry=1;
global time;
time=0;
a = arduino('COM38');
a.pinMode(10, 'output');
% --- Outputs from this function are returned to the command line.
function varargout = DATAloggerMATLAB_OutputFcn(hObject, eventdata,
handles)
varargout{1} = handles.output;
% --- Executes on button press in SOILdata.
function SOILdata_Callback(hObject, eventdata, handles)
global a;
x=0;
while (1)
b=a.analogRead(0);
x=[x,b];
plot(x);
grid on;drawnow;
pause(0.01);
end
% --- Executes on button press in TEMPdata.
function TEMPdata_Callback(hObject, eventdata, handles)
global a;
y=0;
while (1)
c=a.analogRead(1);
```

```
y=[y,c];
plot(y,'r');
grid on;drawnow;
pause(0.01);
end
% --- Executes on button press in LIGHTdata.
function LIGHTdata_Callback(hObject, eventdata, handles)
global a;
z=0;
while (1)
e=a.analogRead(2);
z=[z,e];
plot(z,'g');
grid on;drawnow;
pause(0.01);
end
% --- Executes on button press in DEVICEon.
function DEVICEon_Callback(hObject, eventdata, handles)
global a;
a.digitalWrite(10,1);
% --- Executes on button press in DEVICEoff.
function DEVICEoff_Callback(hObject, eventdata, handles)
global a;
a.digitalWrite(10,0);
% --- Executes on button press in PROCESSstop.
function PROCESSstop_Callback(~, eventdata, handles)
delete(instrfind({'Port'},{'COM38'}));
close all;
```

```
% --- Executes on button press in WATERlevel.
function WATERlevel_Callback(hObject, eventdata, handles)
global a;
w=0;
while (1)
f=a.analogRead(3);
w=[w,f];
plot(w,'g');
grid on;drawnow;
pause(0.01);
end
% --- Executes on button press in HUMdata.
function HUMdata_Callback(hObject, eventdata, handles)
global a;
s=0;
while (1)
h=a.analogRead(4);
s=[s,h];
plot(s,'r');
grid on;drawnow;
pause(0.01);
end
% --- Executes on button press in SMOKEdata.
function SMOKEdata_Callback(hObject, eventdata, handles)
global a;
r=0;
while (1)
i=a.analogRead(5);
```

```
  r=[r,i];
  plot(r,'b');
  grid on;drawnow;
   pause(0.01);
end
```

7.6 MATLAB GUI for the system

Follow the steps described in the 7.3, 7.4 &7.5 sections, MATLAB GUI are designed, Fig.7.3 (GUI showing smoke detector data on graph), 7.4(GUI showing light data on graph).

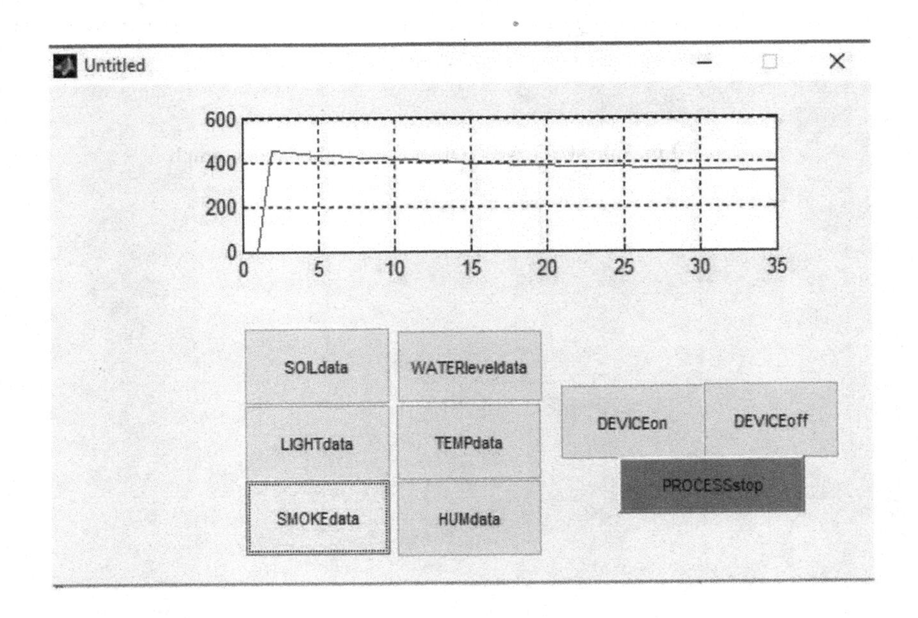

Fig. 7.3: MATLAB GUI showing smoke data on graph

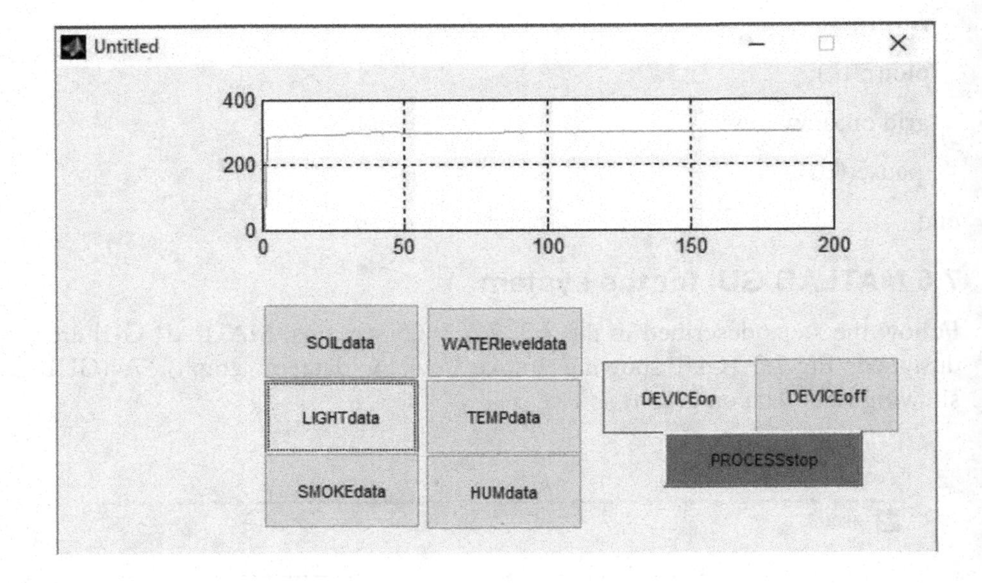

Fig. 7.4: MATLAB GUI showing light data on graph

8

A Smart Control for Site-specific Management of Fixed Irrigation System

8.1 Introduction

The objective of this chapter is to develop a smart control for site-specific management of fixed irrigation system with Blynk app. Mobile app is designed to control the water management in field. The complete system comprises of two sections field device and mobile app. Field device comprises of Arduino, Node MCU, Power supply, LCD, Relay board, soil moisture sensor, temperature & humidity sensor, water level sensor, motor1, motor2. The system is designed to establish control and communication with specific agricultural field to take sensory data from the sensors and control the PUMP IN motor and PUMP OUT motor with the help of mobile app. Fig.8.1 shows a block diagram for a system.

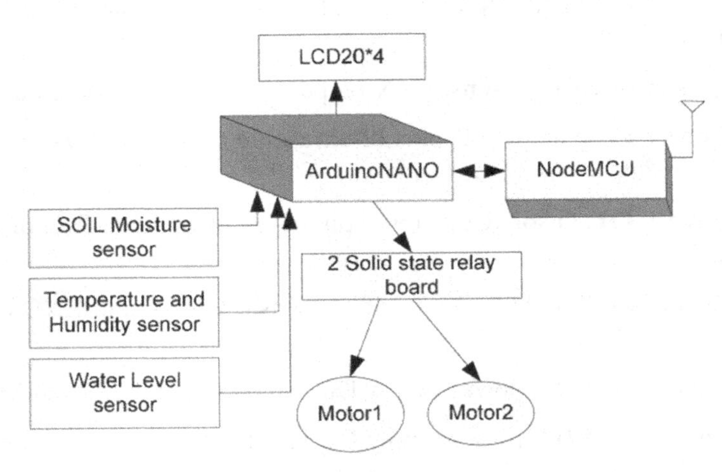

Fig. 8.1: Block diagram of the system

Table 8.1: Components list

Component/Specification	Quantity
Power supply 12V/1Amp	1
2 Relay Board	1
Jumper wire M-M	20
Jumper wire M-F	20
Jumper wire F-F	20
Power supply extension (To get more +5V and GND)	1
LCD20*4	1
LCD patch/explorer board	1
NodeMCU patch	1
NodeMCU	1
Soil moisture sensor-SERIAL OUT	1
Ultrasonic sensor patch	1
Temp and HUM sensor-SERIAL OUT	1
Arduino UNO	1

Note: All components are available at www.nuttyengineer.com

8.2 Circuit Diagram

Connection

1. Connect **SOIL sensor** output pin OUTPUT_SS to pinA0 of Arduino UNO
2. Connect +Vcc and GND pins of SOIL sensor to +5V and GND of Power supply
3. Connect **Ultrasonic sensor** RX-output1 pin to pin RX of Arduino UNO
4. Connect +Vcc and GND pins of ultrasonic sensor to +5V and GND of Power supply
5. Connect **TH sensor** RX-output2 pin to pin 6(mySerial RX) of Arduino UNO
6. Connect +Vcc and GND pins of TH sensor to +5V and GND of Power supply
7. Connect +12V/1A power supply DC jack to DC jack of NodeMCU
8. Connect +12V/1A power supply DC jack to DC jack of Arduino UNO
9. Pins RS, RW and E of LCD is connected to pins D0, GND and D1 of NodeMCU.

10. Pins D4, D5, D6 and D7of LCD are connected to pins D2, D3, D4 and D5 of NodeMCU.
11. Pins 1,3 and 16 of LCD are connected to GND of power supply using power supply patch.
12. Pins 2 and 15 of LCD are connected to +5V of power supply using power supply patch.
13. Water Pump IN motor and Water Pump OUT motor to D6and D7 pins of NodeMCU using relay board.
14. The base of NPN transistor 2N2222 is to be connected with pins of nodeMCU, in this case four pins D6,D7,D8,D9.
15. Emitter of transistor is grounded.
16. Collector of transistor is to be connected with L2 of relay and Li of relay to positive terminal of 12V battery.
17. Negative terminal of battery is connected with ground.
18. One terminal of appliance (pump motor) is connected with 'NO' of relay and other to one end the AC source.
19. Other end of AC source is connected to 'Common' terminal of relay.
20. TX(1) pin of Arduino Uno is connected to D7(myserial RX) pin of Node MCU.

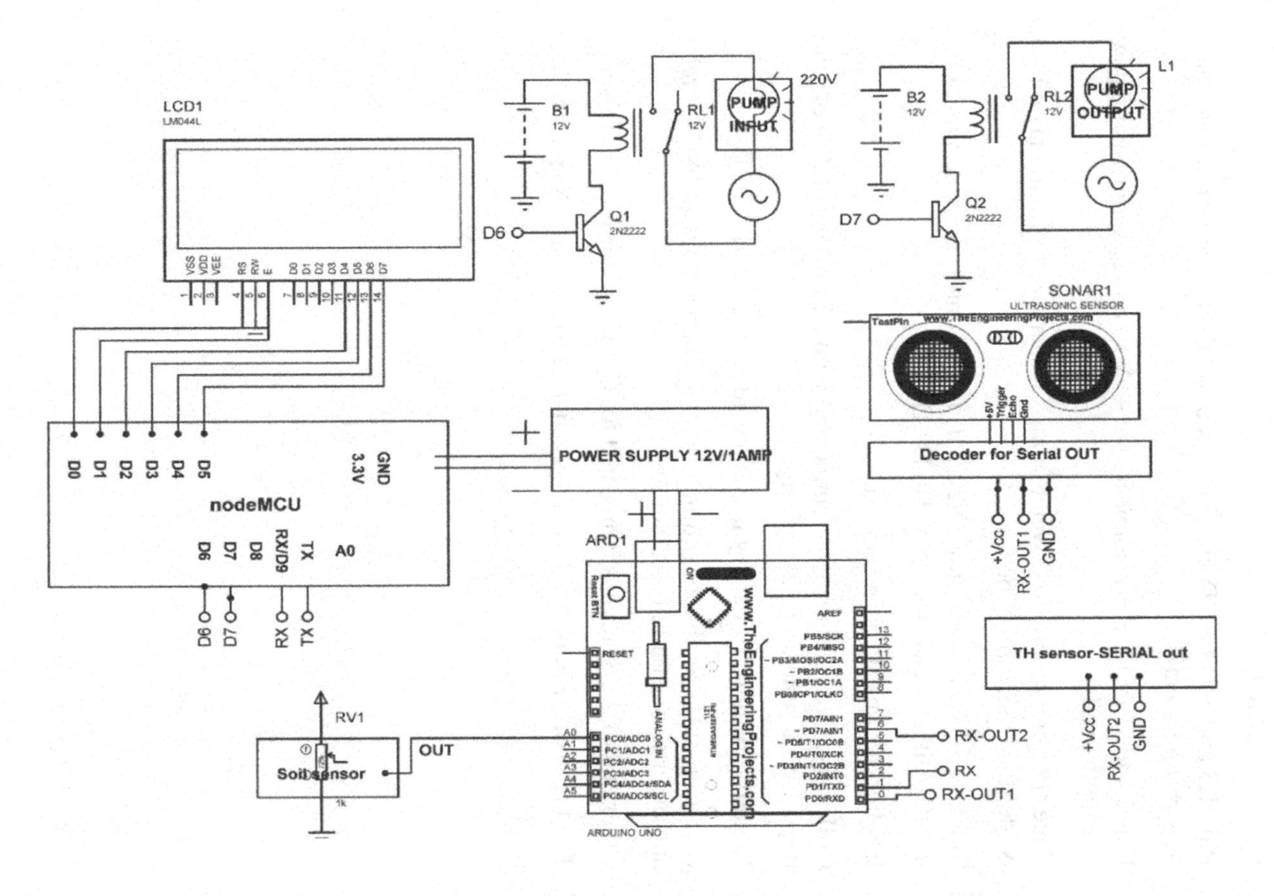

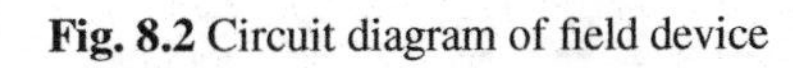
Fig. 8.2 Circuit diagram of field device

8.3 Program

8.3.1 Arduino code

```
#include <LiquidCrystal.h>
LiquidCrystal lcd(13, 12, 11, 10, 9, 8);
#include <SoftwareSerial.h>
SoftwareSerial mySerial(6,7);// 6 rx /7 tx
#define SOIL_SENSOR A0
String inputString_ULTRA = "";
String inputString_TH="";
String ULTRA;
int TEMP,HUM;
void setup()
{
Serial.begin(9600);
mySerial.begin(9600);
lcd.begin(20, 4);
}
void loop()
{
TEMP_HUM_READ();
ULTRASONIC_READ();
int SOIL_value=analogRead(SOIL_SENSOR);//////////soil read
Serial.print(SOIL_value);
Serial.print(",");
Serial.print(ULTRA);
Serial.print(",");
```

```
Serial.print(TEMP);
Serial.print(",");
Serial.print(HUM);
Serial.print('\n');
}
void ULTRASONIC_READ()
{
while (Serial.available()>0)
{
inputString_ULTRA = Serial.readStringUntil('\r');// Get serial input
ULTRA=String(((inputString_ULTRA[0]-48)*100) + ((inputString_UL-
TRA[1]-48)*10)+ ((inputString_ULTRA[2]-48)*1))+"."+String (((input-
String_ULTRA[4]-48)*10)+ ((inputString_ULTRA[5]-48)*1));
}
inputString_ULTRA = "";
delay(20);
}
void TEMP_HUM_READ()
{
while (mySerial.available()>0)
{
 inputString_TH = mySerial.readStringUntil('\r');// Get serial input
 HUM=(((inputString_TH[3]-48)*100) + ((inputString_TH[4]-48)*10)+
((inputString_TH[5]-48)*1));
TEMP=(((inputString_TH[9]-48)*100) + ((inputString_TH[10]-48)*10)+
((inputString_TH[11]-48)*1));
}
inputString_TH = "";
delay(20);
}
```

8.3.2 NodeMCU code

```
#include "StringSplitter.h"
#define BLYNK_PRINT Serial
///// library for external LCD
#include <LiquidCrystal.h>
LiquidCrystal lcd(D0, D1, D2, D3, D4, D5);
////// library for NodeMCU
#include <ESP8266WiFi.h>
#include <BlynkSimpleEsp8266.h>
#include <SoftwareSerial.h>
SoftwareSerial rajSerial(D7,D8,false,256);
char auth[] = "5c8e33bf09a04b03b2fa153928b075c5";///rediff rajesh
char ssid[] = "ESPServer_RAJ";
char pass[] = "RAJ@12345";
//////// library for internal LCD
WidgetLCD LCD_BLYNK(V8);
///// for timer
BlynkTimer timer;
int PUMP_IN=12;//D6
int PUMP_OUT=13;//D7
String ULTRA,TEMP,HUM,SOIL;
String CONT_NEW_STRING= "";
////////////////// use button
BLYNK_WRITE(V1)
{
int PUMP_IN_VAL = param.asInt();
```

```
if(PUMP_IN_VAL==HIGH)
{
lcd.clear();
digitalWrite(PUMP_IN,HIGH);
digitalWrite(PUMP_OUT,LOW);
////// external LCD with NOdeMCU
lcd.setCursor(0,0);
lcd.print("PUMP_In Tigger");
//// LCD blynk
LCD_BLYNK.print(0,0,"PUMP_In Tigger");
delay(10);
}
}
BLYNK_WRITE(V2)
{
int PUMP_OUT_VAL = param.asInt();
if(PUMP_OUT_VAL==HIGH)
{
lcd.clear();
digitalWrite(PUMP_IN,LOW);
digitalWrite(PUMP_OUT,HIGH);
 ////// external LCD with nodeMCU
lcd.setCursor(0,0);
lcd.print("PUMP_OUT Tigger");
//// LCD blynk
LCD_BLYNK.print(0,0,"PUMP_OUT Tigger");
delay(10);
}
```

```
}
BLYNK_WRITE(V3)
{
int BOTH_ON = param.asInt();
if(BOTH_ON==HIGH)
{
lcd.clear();
digitalWrite(PUMP_IN,HIGH);
digitalWrite(PUMP_OUT,HIGH);
////// external LCD with nodeMCU
lcd.setCursor(0,0);
lcd.print("BOTH ON");
//// LCD blynk
LCD_BLYNK.print(0,0,"BOTH ON");
delay(10);
}
}
/////// read analog sensor
void READ_SENSOR()
{
serialEvent_NODEMCU();
Blynk.virtualWrite(V4,SOIL);
Blynk.virtualWrite(V5,ULTRA);
Blynk.virtualWrite(V6,TEMP);
Blynk.virtualWrite(V7,HUM);
lcd.setCursor(0,1);
lcd.print("SOIL:");
lcd.setCursor(5,1);
```

```
lcd.print(SOIL);
lcd.setCursor(0,2);
lcd.print("LEVEL:");
lcd.setCursor(6,2);
lcd.print(ULTRA);
lcd.setCursor(0,3);
lcd.print("TEMP:");
lcd.setCursor(5,3);
lcd.print(TEMP);
lcd.setCursor(10,3);
lcd.print("HUM:");
lcd.setCursor(15,3);
lcd.print(HUM);
}
void setup()
{
Serial.begin(9600);
lcd.begin(20, 4);
Blynk.begin(auth, ssid, pass);
pinMode(PUMP_IN,OUTPUT);//D6 pin of NodeMCU
pinMode(PUMP_OUT,OUTPUT);//D7 pin of NodeMCU
timer.setInterval(1000L,READ_SENSOR);//// read sensor with setting delay of 1 Sec
}
void loop()
{
Blynk.run();
timer.run(); // Initiates BlynkTimer
}
```

```
void serialEvent_NODEMCU()
{
while (rajSerial.available()>0)
{
CONT_NEW_STRING = rajSerial.readStringUntil('\n');// Get serial input
StringSplitter *splitter = new StringSplitter(CONT_NEW_STRING, ',', 4); // new StringSplitter(string_to_split, delimiter, limit)
int itemCount = splitter->getItemCount();
for(int i = 0; i < itemCount; i++)
{
String item = splitter->getItemAtIndex(i);
SOIL = splitter->getItemAtIndex(0);
ULTRA = splitter->getItemAtIndex(1);
TEMP = splitter->getItemAtIndex(2);
HUM= splitter->getItemAtIndex(3);
}
CONT_NEW_STRING= "";
delay(20);
}
}
```

8.4 BLYNK app

Follow the steps to design Blynk app as discussed in chapter 4. Fig.8.3 Developed BLYNK app (a) pump in 'ON' (b) pump out 'ON'.

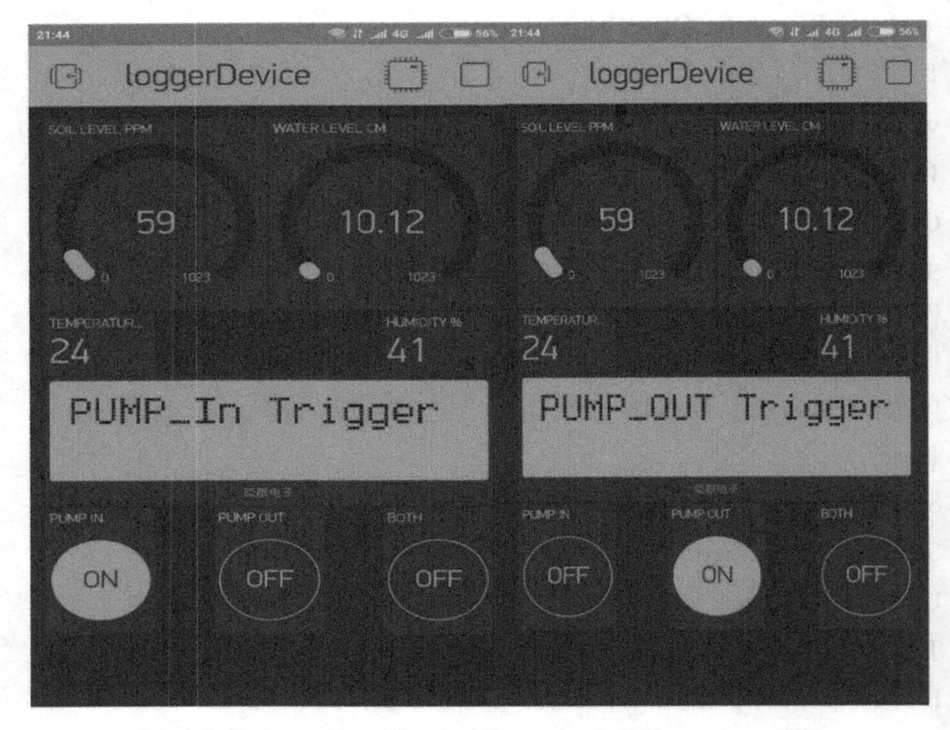

Fig. 8.3: Developed BLYNK app (a) pump in 'ON' (b) pump out 'ON'

9

iCould and Cayenne Based Perimeter Monitoring System for Agricultural Field

9.1 Introduction

The objective of this chapter is to design an icloud and cayenne based perimeter monitoring system for agricultural field. Mobile app is designed to monitor the sensory data. The complete system comprises of two sections field device and mobile app. Field device comprises of NodeMCU, battery, motion sensors and laser sensors. The system is designed to monitor the agriculture field perimeter with motion and laser sensors, which are deployed at different locations of field. The sensory data is transmitted through NodeMCU and received at mobile app which can be accessed by authorized person. The same data can be accessed on webserver from anywhere in the world. Fig.9.1 shows a block diagram for a system. Table.9.1 shows the component list.

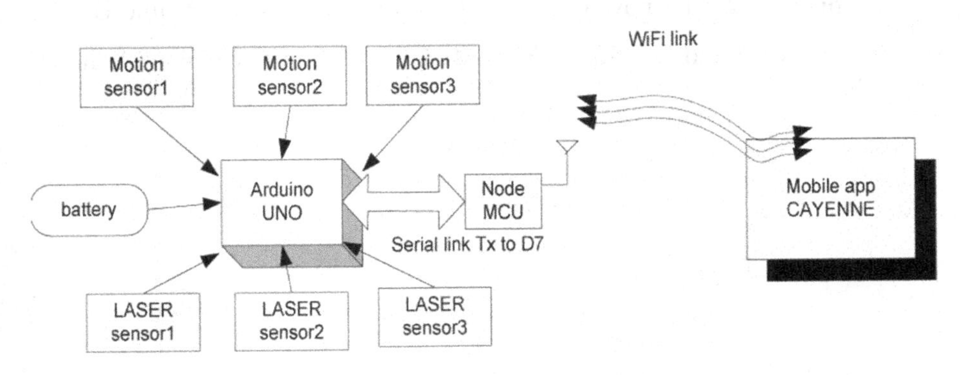

Fig. 9.1: Block diagram of the system

Table:9.1 Components list

Component	Quantity
Power supply 12V/1Amp	1
Arduino uno	1
NodeMCU	1
Jumper wire M-M	20
Jumper wire M-F	20
Jumper wire F-F	20
Power supply extension (To get more +5V and GND)	1
PIR sensor	3
LASER sensor	3
5 Push button array	1
NodeMCU explorer board	1

Note: All components are available at www.nuttyengineer.com

9.2 Circuit Diagram

1. Connect set of three PIR Motion sensor output pins to pin 6,7 and 8 of Arduino UNO
2. Connect +Vcc and GND pins of sensors to +5V and GND of power supply
3. Connect set of three LASER sensor output pin to pins 3, 4 and 5 of Arduino UNO
4. Connect +Vcc and GND pins of RAIN sensors to +5V and GND of Power supply
5. Connect +12V/1A power supply DC jack to DC jack of Arduino UNO
6. Connect D7 and D8 pins of NodeMCU to TX and RX pins of Arduino Uno

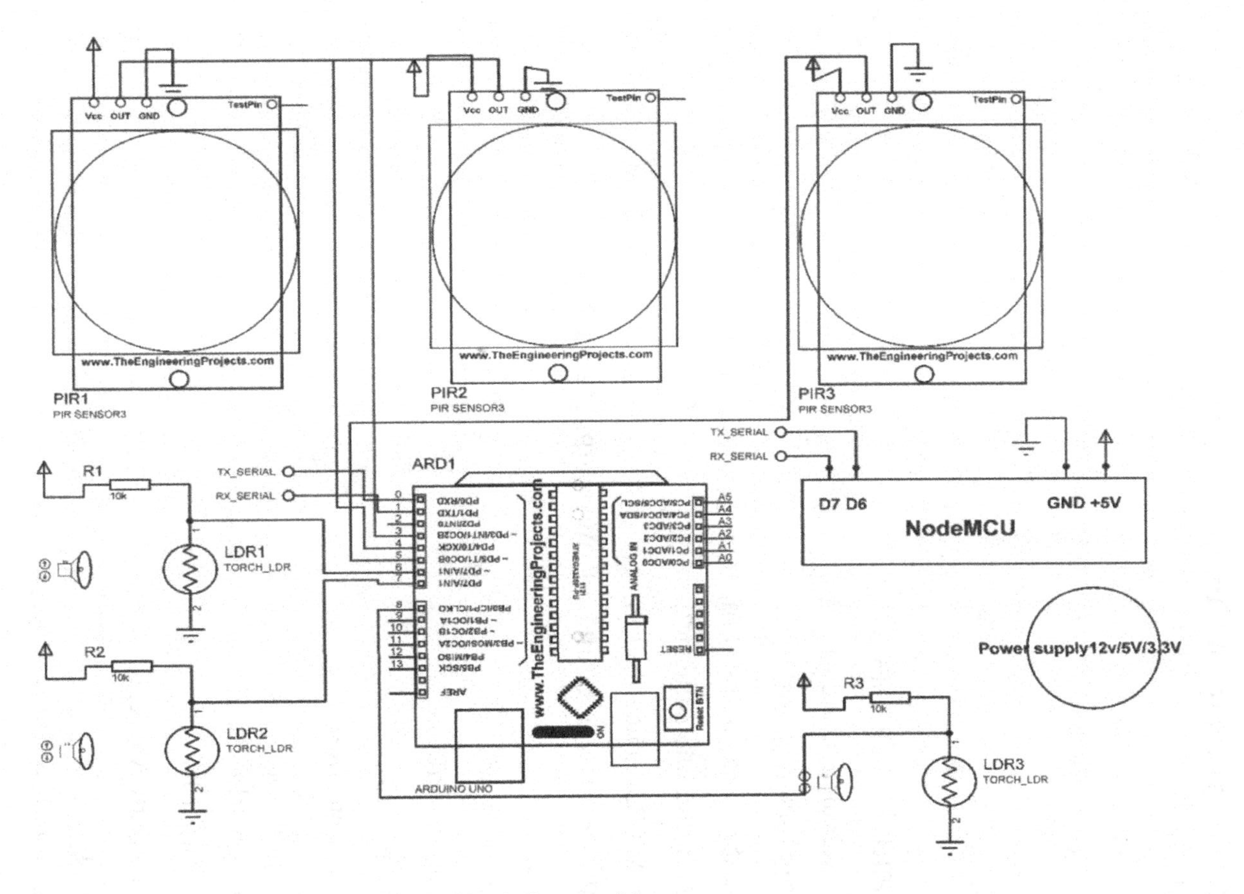

Fig.9.2: Circuit diagram of the system

9.3 Program

9.3.1 Program for Arduino

```
void setup()
{
Serial.begin (9600);
}
void loop()
{
int PIR1=digitalRead(3);
int PIR2=digitalRead(4);
int PIR3=digitalRead(5);
int LASER1=digitalRead(6);
int LASER2=digitalRead(7);
int LASER3=digitalRead(8);
if((PIR1==LOW)&&(PIR2==LOW)&&(PIR3==LOW))
{
Serial.print('\r');
Serial.print(PIR1);
Serial.print('|');
Serial.print(PIR2);
Serial.print('|');
Serial.print(PIR3);
Serial.print('|');
Serial.print(LASER1);
Serial.print('|');
Serial.print(LASER2);
Serial.print('|');
Serial.print(LASER3);
```

```
Serial.print('\n');
delay(30);
}
else if((LASER1==LOW)&&(LASER2==LOW)&&(LASER3==LOW))
{
Serial.print('\r');
Serial.print(PIR1);
Serial.print('|');
Serial.print(PIR2);
Serial.print('|');
Serial.print(PIR3);
Serial.print('|');
Serial.print(LASER1);
Serial.print('|');
Serial.print(LASER2);
Serial.print('|');
Serial.print(LASER3);
Serial.print('\n');
delay(30);
}
else if((PIR1==LOW)&&(PIR2==LOW)&&(PIR3==LOW)&&(LA-
SER1==LOW)&&(LASER2==LOW)&&(LASER3==LOW))
{
Serial.print('\r');
Serial.print(PIR1);
Serial.print('|');
Serial.print(PIR2);
Serial.print('|');
Serial.print(PIR3);
```

```
Serial.print('|');
Serial.print(LASER1);
Serial.print('|');
Serial.print(LASER2);
Serial.print('|');
Serial.print(LASER3);
Serial.print('\n');
delay(30);
}
else
{
Serial.print('\r');
Serial.print(PIR1);
Serial.print('|');
Serial.print(PIR2);
Serial.print('|');
Serial.print(PIR3);
Serial.print('|');
Serial.print(LASER1);
Serial.print('|');
Serial.print(LASER2);
Serial.print('|');
Serial.print(LASER3);
Serial.print('\n');
delay(30);
}
delay(3000);
}
```

9.3.2 Program for NodeMCU

```
//#define CAYENNE_DEBUG
#define CAYENNE_PRINT Serial
#include <CayenneMQTTESP8266.h>
#include <ESP8266WiFi.h>
#include <SoftwareSerial.h>
SoftwareSerial mySerial(D7,D8,false,256);
// WiFi network info.
char ssid[] = "ESPServer_RAJ";
char wifiPassword[] = "RAJ@12345";
// Cayenne authentication info. This should be obtained from the Cayenne
Dashboard.
char username[] = "fac81bb0-7283-11e7-85a3-9540e9f7b5aa";//different for
cash project
char password[] = "3745eb389f4e035711428158f7cdc1adc0475946";
char clientID[] = "386b86f0-7284-11e7-b0bc-87cd67a1f8c7";
int PIR1,PIR2,PIR3,LASER1,LASER2,LASER3;
void setup()
{
pinMode(D0, OUTPUT);
Serial.begin(9600);
Cayenne.begin(username, password, clientID, ssid, wifiPassword);
}
void loop()
{
Cayenne.loop();
SerialDATA();
Cayenne.virtualWrite(0, PIR1);
Cayenne.virtualWrite(1, PIR2);
```

```
Cayenne.virtualWrite(2, PIR3);
Cayenne.virtualWrite(3, LASER1);
Cayenne.virtualWrite(4, LASER2);
Cayenne.virtualWrite(5, LASER3);
delay(500);
}
//Default function for processing actuator commands from the Cayenne Dashboard.
//You can also use functions for specific channels, e.g CAYENNE_IN(1) for channel 1 commands.
CAYENNE_IN_DEFAULT()
{
//CAYENNE_LOG("CAYENNE_IN_DEFAULT(%u) - %s, %s", request.channel, getValue.getId(), getValue.asString());
CAYENNE_LOG("CAYENNE_IN_(1)(%u) - %s, %s", request.channel, getValue.getId(), getValue.asString());
//Process message here. If there is an error set an error message using getValue.setError(), e.g getValue.setError("Error message");
int i = getValue.asInt();
if(i>=45)
{
digitalWrite(D0,HIGH);
}
else
{
digitalWrite(D0,LOW);
}
}
void SerialDATA()
{
```

```
if (mySerial.available()<1)  return;
char STRING_SERIAL=mySerial.read();
if(STRING_SERIAL!='\r') return;
PIR1 =mySerial.parseInt();
PIR2=mySerial.parseInt();
PIR3=mySerial.parseInt();
LASER1=mySerial.parseInt();
LASER2=mySerial.parseInt();
LASER3=mySerial.parseInt();
Serial.print(PIR1);
Serial.print(PIR2);
Serial.print(PIR3);
Serial.print(LASER1);
Serial.print(LASER2);
Serial.println(LASER3);
}
```

9.4 Cayenne App

To design mobile app with cayenne, first device needs to be added to its cloud.

9.4.1 Steps to add NodeMCU in cayenne cloud

1. Install the Arduino IDE and add Cayenne MQTT Library to Arduino IDE.
2. Install the ESP8266 board package to Arduino IDE.
3. Install required USB driver on computer to program the ESP8266.
4. Connect the ESP8266 to PC/Mac via data-capable USB cable.
5. In the Arduino IDE, go to the **tools** menu, select the **board**, and now the **port** ESP8266 is connected to.
6. Use the MQTT username, MQTT password, client ID as well as ssid[] and wifiPassord[] in the arduino IDE to write code

```
//#define CAYENNE_DEBUG
#define CAYENNE_PRINT Serial
#include <CayenneMQTTESP8266.h>
#include <ESP8266WiFi.h>
#include <SoftwareSerial.h>
SoftwareSerial mySerial(D7,D8,false,256);
// WiFi network info.
char ssid[] = "ESPServer_RAJ";
char wifiPassword[] = "RAJ@12345";

// Cayenne authentication info. This should be obtained from the Cayenne Dashboard.
char username[] = "fac81bb0-7283-11e7-85a3-9540e9f7b5aa";
char password[] = "3745eb389f4e035711428158f7cdc1adc0475946";
char clientID[] = "386b86f0-7284-11e7-b0bc-87cd67a1f8c7";
```

Fig. 9.3: Snapshot showing username, password to MQTT

7. Burn the code in Arduino and NodeMCU then window will open, Fig.9.4. Fig.9.5 shows the snapshots for the developed mobile app after burning program.

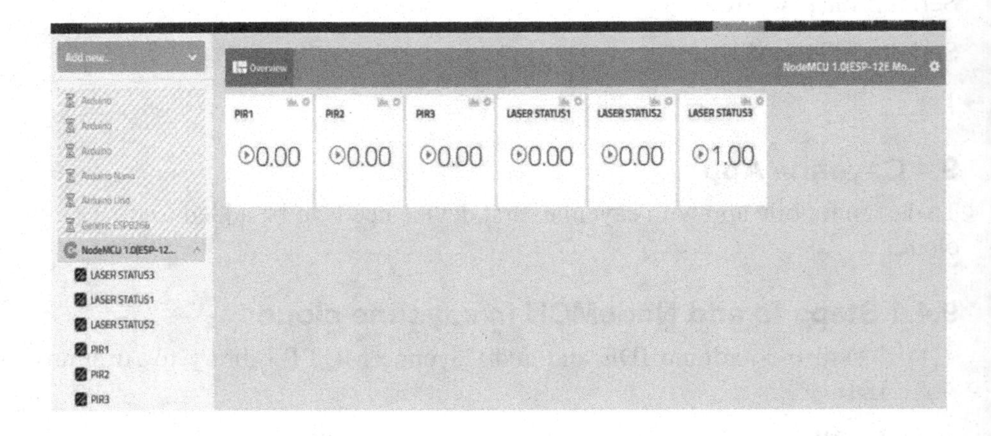

Fig.9.4: Sensory data received at Cayenne webserver

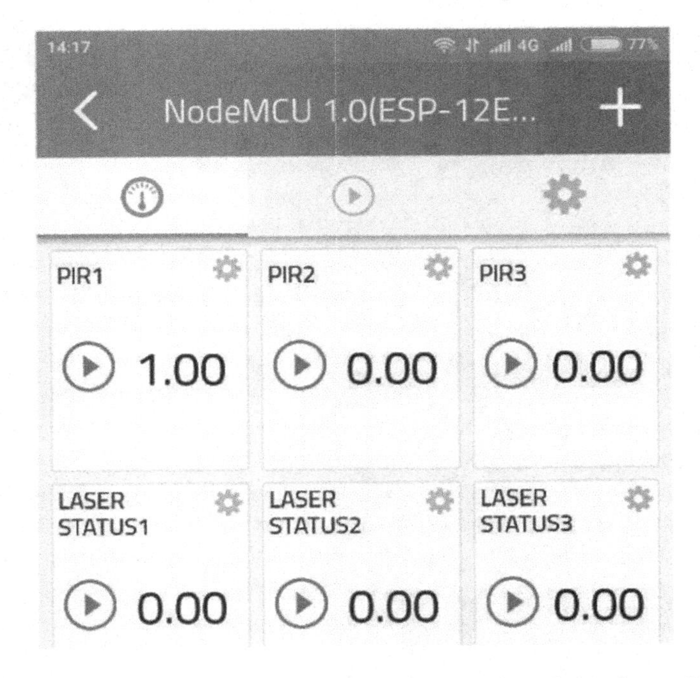

Fig.9.5: Snapshot of Mobile app (PIR1 shows human being detected)

10

Scilab Based Data Logger for Plant Protection from Fire in Agricultural Field

10.1 Introduction

The objective of this chapter is to develop a data logger and indicator for fire detection in agriculture field. The data logger is designed with the help of Scilab. The system comprises of Ardiuini Nano, light sensor, MQ6, MQ135, IR sensor, temperature sensor, power supply. The system is designed to take sensory data on Scilab with serial communication. Fig.10.1 shows block diagram for a system. Table.10.1 shows the component list required to develop the system.

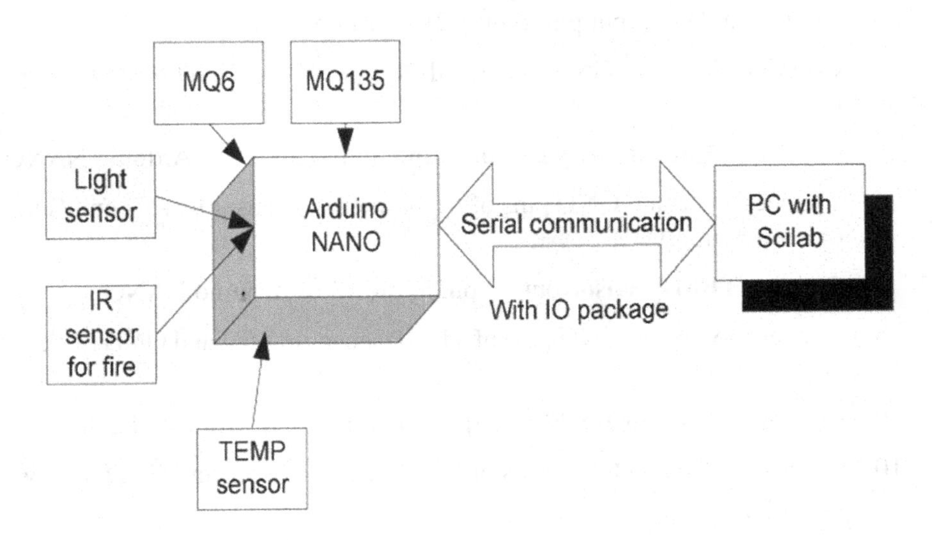

Fig.10.1: Block diagram of the system

Table:10.1 Components list

Component	Quantity
Power supply 12V/1.5Amp	1
Arduino UNO	1
Jumper wire M-M	20
Jumper wire M-F	20
Jumper wire F-F	20
Power supply extension (To get more +5V and GND)	1
MQ135	1
MQ6	1
Light sensor	1
Temperature sensor	1
Touch sensor	1

Note: All components are available at www.nuttyengineer.com

10.2 Circuit Diagram

Connections

1. Connect **MQ135** output pin to pinA1 of Arduino NANO
2. Connect +Vcc and GND pins of **MQ135** sensors to +5V and GND of Power supply
3. Connect **MQ6** output pin to pinA2 of Arduino NANO
4. Connect +Vcc and GND pins of **MQ6** to +5V and GND of Power supply
5. Connect **Light intensity sensor** output pin to pinA0 of Arduino NANO
6. Connect +Vcc and GND pins of Light intensity sensor to +5V and GND of Power supply
7. Connect **TEMP sensor** output pin to pinA3 of Arduino NANO
8. Connect +Vcc and GND pins of TEMP sensor to +5V and GND of Power supply
9. Connect **IR sensor for Fire** output pin to pin 6 of Arduino NANO
10. Connect +Vcc and GND pins of HUM sensor to +5V and GND of Power supply
11. Connect **touch sensor** output pin to pin 7 of Arduino NANO

12. Connect +Vcc and GND pins of touch sensor to +5V and GND of Power supply
13. Connect +12V/1.5A battery DC jack to DC jack of Arduino NANO.
14. Connect two LEDs as indicator for IR and touch sensor at pin5 and 4 of Arduino NANO through 330 ohm of resistors

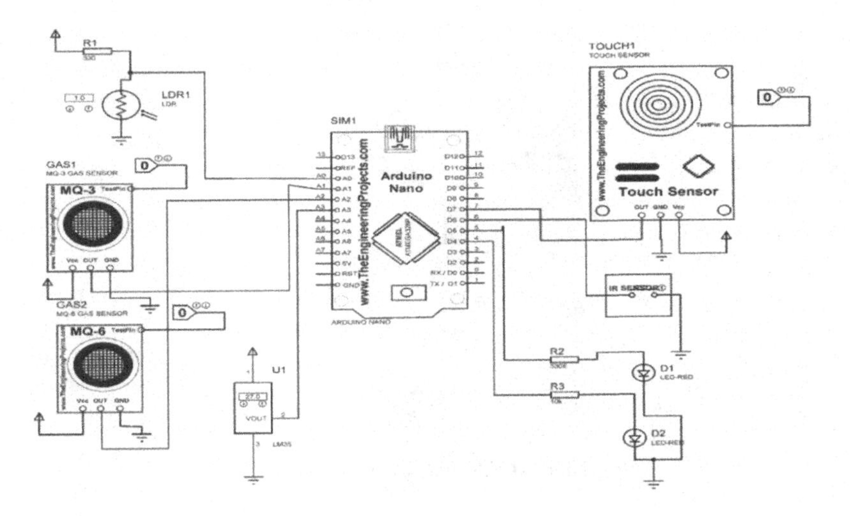

Fig.10.2: Circuit diagram for the system

10.3 Procedure to run Arduino I/O package with Scilab

1. Download Arduino I/O package for Scilab
2. Open the package and install the firmware from the file Arduino-1.1 to the Scilab platform
3. Again open the Arduino I/O folder and open the pde folder and upload the toolbox_arduino_v3_ino in the arduino microcontroller
4. After uploading the program, connect the GUI to scilab to read the data

10.4 Scilab XCOS model

Steps to design Scilab XCOS model

1. Open Scilab window
2. Open XCOS windoe
3. Go to view and then click on "Palette browser"

4. Two windows will be opened. One with blocks and other 'blank'
5. Make right click on the block and click on add.
6. Block will appear in black window.
7. Complete the XCOS model.

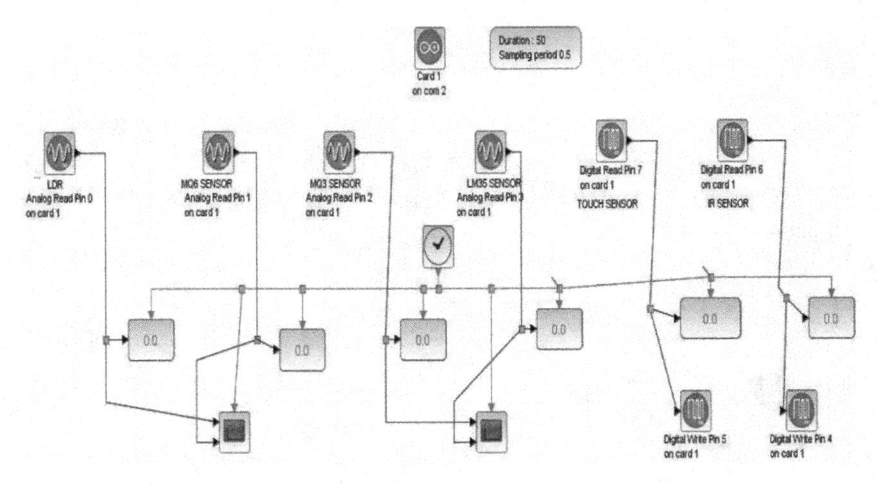

Fig.10.3: XCOS model for the system

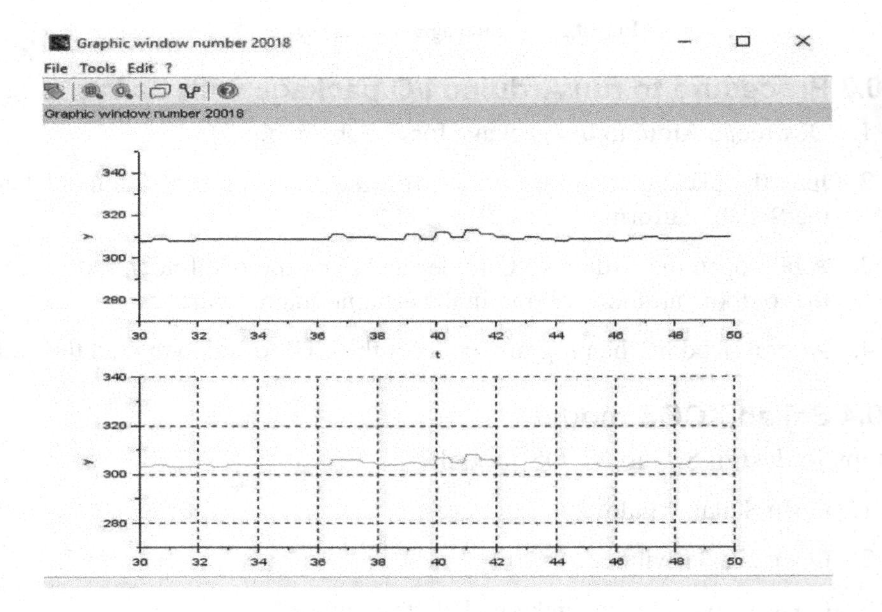

Fig.10.4: Waveform Chart snap1

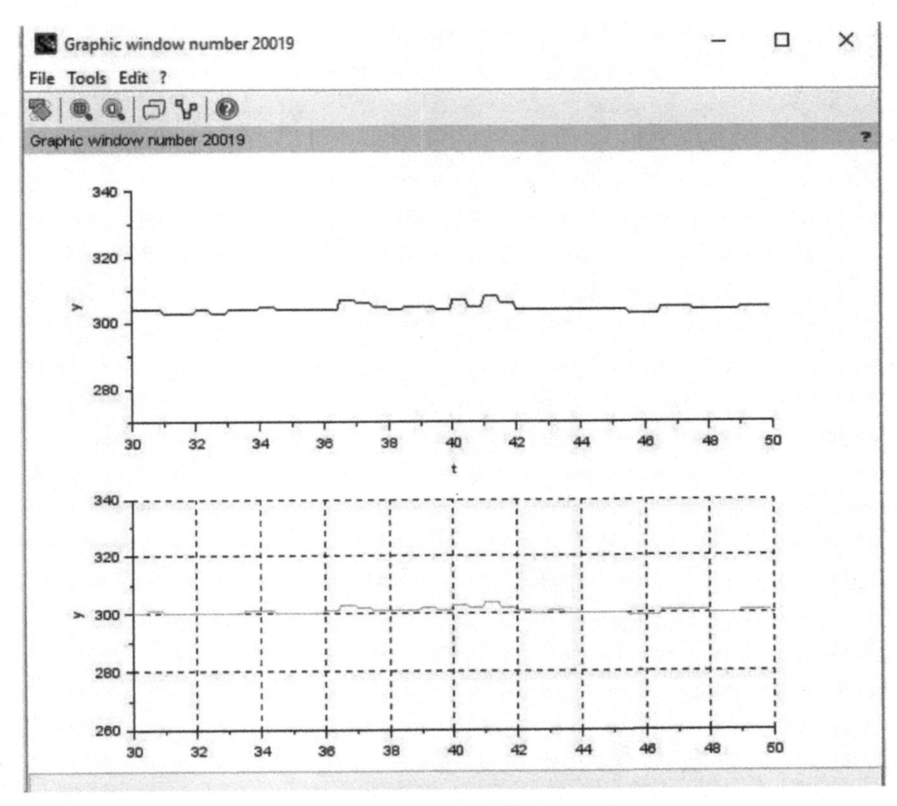

Fig.10.5: Waveform Chart snap2

11

Smart Greenhouse Monitoring System for Flower Plant Growth

11.1 Introduction

The application of greenhouse in agriculture is vast. Greenhouses are used for growing flowers, vegetables, fruits, transplants and varieties of certain crops in a controlled environment which is appropriate for the plants. The edible flowers are served in some good restaurants to grow them precise calculation is required. Sensor network can help to monitor and automate the activities in greenhouse. The objective of this chapter is to discuss the development of sensor nodes to monitor the growth of flower plants through XBee and internet of things. The system comprises of three sections- sensor nodes, local server and main server. Sensor nodes communicate to local server with XBee and then local server communicates the same data to main server through IoT. Fig.11.1 shows the generalized block diagram, sensor node is to be implemented on each flower pot and communicating to local and main server.

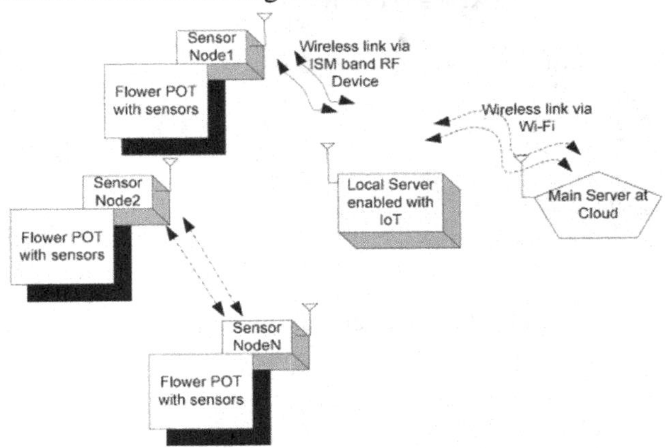

Fig.11.1: Generalized block diagram of the system

Fig.11.2: shows the block diagram of sensor node which comprises of Arduino, battery, water pump, light sensor, soil moisture sensor, temperature/humidity sensor, rainfall sensor, altitude/pressure sensor, XBee.

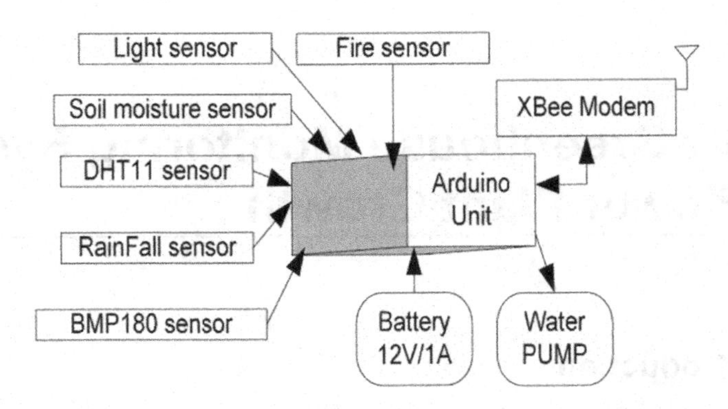

Fig.11.2 Block diagram of sensor Node

Note: Sensors can be added or removed as per requirement.

Fig.11.3: shows the block diagram of local server and PC as main server. Local server comprises of Arduino, battery, LCD, XBee, WiFi modem.

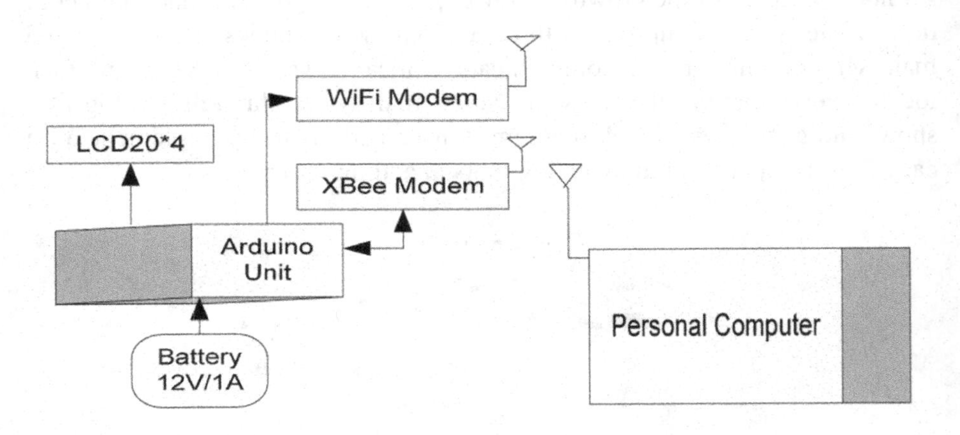

Fig.11.3: Block diagram of sensor Node

Table:11.1: Components list for sensor node

Component	Quantity
Power supply 12V/1Amp	1
Arduino nano	1
XBee modem	1
Jumper wire M-M	20
Jumper wire M-F	20
Jumper wire F-F	20
Power supply extension (To get more +5V and GND)	1
Rainfall sensor	1
Fire sensor	1
Soil sensor	1
Light sensor	1
DHT11	1
BMP180	1
5 Push button array	1
XBee explorer board	1

Table 11.2: Components list for Local Server

Component/Specification	Quantity
Power supply 12V/1Amp	1
Arduino Nano	1
XBee modem	1
Jumper wire M-M	20
Jumper wire M-F	20
Jumper wire F-F	20
Power supply extension (To get more +5V and GND)	1
LCD20*4	1
LCD patch/explorer board	1
5 Push button array	1
XBee explorer board	1
NodeMCU patch	1
NodeMCU	1

Note: All components are available at www.nuttyengineer.com

11.2 Circuit Diagram

11.2.1 Sensor Node

1. Connect **Fire sensor** output pin OUTPUT_FS to pin13 of Arduino Nano
2. Connect +Vcc and GND pins of sensors to +5V and GND of power supply
3. Connect **RAIN sensor** output pin OUTPUT_RS to pinA0 of Arduino Nano
4. Connect +Vcc and GND pins of RAIN sensors to +5V and GND of Power supply
5. Connect **SOIL sensor** output pin OUTPUT_SS to pinA1 of Arduino Nano
6. Connect +Vcc and GND pins of SOIL sensor to +5V and GND of Power supply
7. Connect **Light sensor** output pin OUTPUT_LS to pinA2 of Arduino Nano
8. Connect +Vcc and GND pins of Light sensor to +5V and GND of Power supply
9. Connect pin 2 of **DHT11** to pin 2 of Arduino Nano
10. Connect +Vcc and GND pins of DHT11 sensor to +5V and GND of Power supply
11. Connect SDA and SCL pins of **BMP180** sensor to A4 and A5 pins of Arduino Nano
12. Connect +Vcc and GND pins of DHT11 sensor to +5V and GND of Power supply
13. Connect TX, RX, +Vcc and GND pins of XBee to pins 6, 7, +5V and GND of Arduino Nano.
14. Connect +12V/1A power supply DC jack to DC jack of Arduino Nano.
15. Water Pump and exhaust Fan to D5and D4 pins of Arduino NANO using relay board
16. The base of NPN transistor 2N2222 is to be connected with pins of nodeMCU, in this case four pins D6,D7,D8,D9.
17. Emitter of transistor is grounded.

18. Collector of transistor is to be connected with L2 of relay and Li of relay to positive terminal of 12V battery.
19. Negative terminal of battery is connected with ground.
20. One terminal of appliance (pump motor/exhaust) is connected with 'NO' of relay and other to one end the AC source.
21. Other end of AC source is connected to 'Common' terminal of relay.

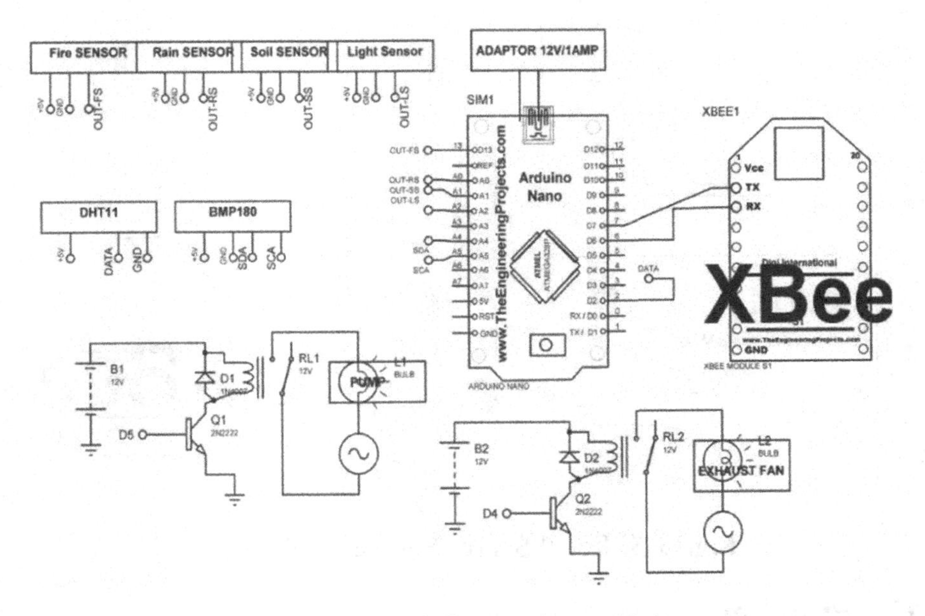

Fig.11.2: Block diagram of sensor Node

11.2.2 Local Server

Connections

1. Connect TX, RX, +Vcc and GND pins of XBee to pins 6, 7, +5V and GND of Arduino Nano.
2. Connect +12V/1A power supply DC jack to DC jack of Arduino Nano.
3. Connect D7 and D8 pins of NodeMCU to TX and RX pins of Arduino Nano
4. Pins RS, RW and E of LCD is connected to pins D0, GND and D1 of NodeMCU.
5. Pins D4, D5, D6 and D7of LCD are connected to pins D2, D3, D4 and D5 of NodeMCU.

6. Pins 1,3 and 16 of LCD are connected to GND of power supply using power supply patch.
7. Pins 2 and 15 of LCD are connected to +5V of power supply using power supply patch.

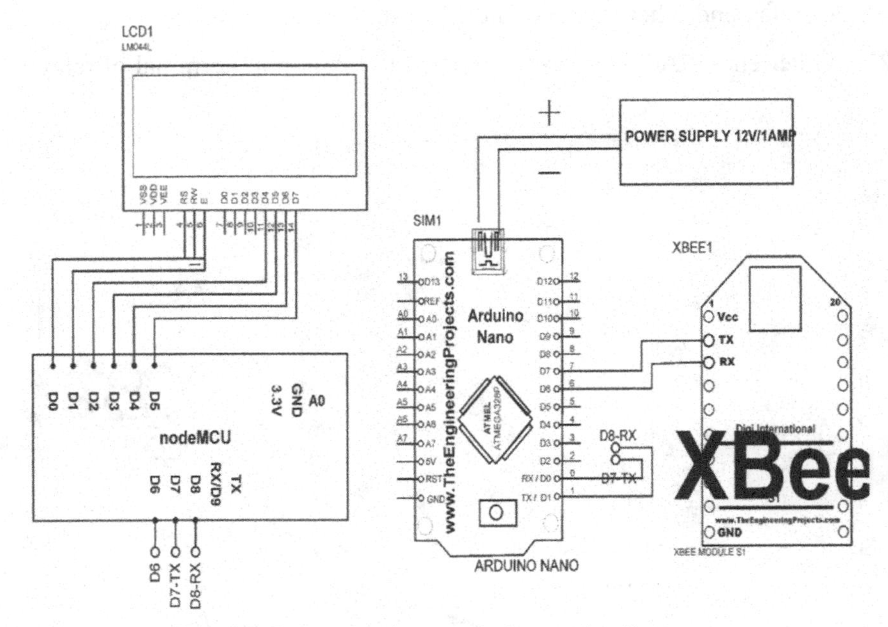

Fig.11.5: Local Server Schematic diagram

11.3 Program

11.3.1 Sensor Node

```
///// library for BMP185

#include <Wire.h>

#include <Adafruit_BMP085.h>

Adafruit_BMP085 bmp;

//////// library for DHT11

#include <dht.h>

dht DHT;

#define DHT11_PIN 2
```

```
/////////////// library for LCD
#include <LiquidCrystal.h>
LiquidCrystal lcd(13, 12, 11, 10, 9, 8);
////////////// libray for Softserial
#include <SoftwareSerial.h>
SoftwareSerial mySerial(6,7);// 6 rx /7 tx
int Fire_level,SOIL_level,LIGHT_level,RAIN_level;
void setup()
{
Serial.begin(9600);
mySerial.begin(9600);
lcd.begin(20, 4);
bmp.begin();
}
void loop()
{
lcd.clear();
//////// read Fire sensor
Fire_level=digitalRead(13);
//////// read Soil sensor
SOIL_level=analogRead(A0);
SOIL_level=SOIL_level/2;
//////// read light sensor
LIGHT_level=analogRead(A1);
//////// read RAIN sensor
RAIN_level=analogRead(A2);
//// read DHT sensor
int chk = DHT.read11(DHT11_PIN);
```

```
if(Fire_level==LOW)
{
int FIRE_level=10;
/////////// soil sensor
lcd.setCursor(0,0);
lcd.print("SOIL:");
lcd.print(SOIL_level);
////// read air quality sensor
lcd.setCursor(10,0);
lcd.print("LIGT:");
lcd.print(LIGHT_level);
//////// read rain sensor level
lcd.setCursor(0,1);
lcd.print("RAIN:");
lcd.print(RAIN_level);
/////// fire
lcd.setCursor(10,1);
lcd.print("FStatus:");
lcd.print("Y");
////// read and Display DHT
lcd.setCursor(0,2);
lcd.print("T:");
lcd.print(DHT.temperature);
lcd.setCursor(10,2);
lcd.print("H:");
lcd.print(DHT.humidity);
////////////////////////// read and display BMP185 data
lcd.setCursor(0,3);
```

```
lcd.print("P0:");
lcd.print(bmp.readPressure());
lcd.print("Pa");
lcd.setCursor(10,3); // Calculate altitude assuming 'standard' barometric &
pressure of 1013.25 millibar = 101325 Pascal
lcd.print("A0:");
lcd.print(bmp.readAltitude());
lcd.print("m");
Serial.print(SOIL_level);
Serial.print(",");
Serial.print(LIGHT_level);
Serial.print(",");
Serial.print(RAIN_level);
Serial.print(",");
Serial.print(FIRE_level);
Serial.print(",");
Serial.print(DHT.temperature);
Serial.print(",");
Serial.print(DHT.humidity);
Serial.print(",");
Serial.print(bmp.readAltitude());
Serial.print(",");
Serial.print(bmp.readPressure());
Serial.print('\n');
delay(30);
mySerial.print(SOIL_level);
mySerial.print(",");
mySerial.print(LIGHT_level);
mySerial.print(",");
```

```
mySerial.print(RAIN_level);
mySerial.print(",");
mySerial.print(FIRE_level);
mySerial.print(",");
mySerial.print(DHT.temperature);
mySerial.print(",");
mySerial.print(DHT.humidity);
mySerial.print(",");
mySerial.print(bmp.readAltitude());
mySerial.print(",");
mySerial.print(bmp.readPressure());
mySerial.print('\n');
delay(30);
}
else
{
 int FIRE_level=20;
 /////////// soil sensor
lcd.setCursor(0,0);
lcd.print("SOIL:");
lcd.print(SOIL_level);
 ////// read air quality sensor
lcd.setCursor(10,0);
lcd.print("LIGT:");
lcd.print(LIGHT_level);
//////// read rain sensor level
lcd.setCursor(0,1);
lcd.print("RAIN:");
```

```
lcd.print(RAIN_level);
/////// fire
lcd.setCursor(10,1);
lcd.print("FStatus:");
lcd.print("Y");
////// read and Display DHT
lcd.setCursor(0,2);
lcd.print("T:");
lcd.print(DHT.temperature);
lcd.setCursor(10,2);
lcd.print("H:");
lcd.print(DHT.humidity);
///////////////////////////// read and display BMP185 data
lcd.setCursor(0,3);
lcd.print("P0:");
lcd.print(bmp.readPressure());
lcd.print("Pa");
lcd.setCursor(10,3); // Calculate altitude assuming 'standard' barometric &
pressure of 1013.25 millibar = 101325 Pascal
lcd.print("A0:");
lcd.print(bmp.readAltitude());
lcd.print("m");
Serial.print(SOIL_level);
Serial.print(",");
Serial.print(LIGHT_level);
Serial.print(",");
Serial.print(RAIN_level);
Serial.print(",");
Serial.print(FIRE_level);
```

```
Serial.print(",");
Serial.print(DHT.temperature);
Serial.print(",");
Serial.print(DHT.humidity);
Serial.print(",");
Serial.print(bmp.readAltitude());
Serial.print(",");
Serial.print(bmp.readPressure());
Serial.print('\n');
delay(30);
mySerial.print(SOIL_level);
mySerial.print(",");
mySerial.print(LIGHT_level);
mySerial.print(",");
mySerial.print(RAIN_level);
mySerial.print(",");
mySerial.print(FIRE_level);
mySerial.print(",");
mySerial.print(DHT.temperature);
mySerial.print(",");
mySerial.print(DHT.humidity);
mySerial.print(",");
mySerial.print(bmp.readAltitude());
mySerial.print(",");
mySerial.print(bmp.readPressure());
mySerial.print('\n');
delay(30);
}
}
```

11.3.2 Local Server

```
#include "ThingSpeak.h"
#include <SoftwareSerial.h>
SoftwareSerial rajSerial(D7,D8,false,256);
#include "StringSplitter.h"
#if !defined(USE_WIFI101_SHIELD) && !defined(USE_ETHERNET_
SHIELD) && !defined(ARDUINO_SAMD_MKR1000) &&
!defined(ARDUINO_AVR_YUN) && !defined(ARDUINO_ARCH_
ESP8266)
  #error "Uncomment the #define for either USE_WIFI101_SHIELD or
USE_ETHERNET_SHIELD"
#endif
#if defined(ARDUINO_AVR_YUN)
#include "YunClient.h"
YunClient client;
#else
  #if defined(USE_WIFI101_SHIELD) || defined(ARDUINO_SAMD_
MKR1000) || defined(ARDUINO_ARCH_ESP8266)
// Use WiFi
#ifdef ARDUINO_ARCH_ESP8266
#include <ESP8266WiFi.h>
#else
#include <SPI.h>
#include <WiFi101.h>
#endif
char ssid[] = " RAJESH";    // WiFi network name
char pass[] = "12345";   // network password
int status = WL_IDLE_STATUS;
   WiFiClient  client;
  #elif defined(USE_ETHERNET_SHIELD)
```

```
    // Use wired ethernet shield
    #include <SPI.h>
    #include <Ethernet.h>
    byte mac[] = { 0xDE, 0xAD, 0xBE, 0xEF, 0xFE, 0xED};
    EthernetClient client;
  #endif
#endif
#ifdef ARDUINO_ARCH_AVR
  // On Arduino:  0 - 1023 maps to 0 - 5 volts
  #define VOLTAGE_MAX 5.0
  #define VOLTAGE_MAXCOUNTS 1023.0

#elif ARDUINO_SAMD_MKR1000
  // On MKR1000:  0 - 1023 maps to 0 - 3.3 volts
  #define VOLTAGE_MAX 3.3
  #define VOLTAGE_MAXCOUNTS 1023.0
#elif ARDUINO_SAM_DUE
  // On Due:  0 - 1023 maps to 0 - 3.3 volts
  #define VOLTAGE_MAX 3.3
  #define VOLTAGE_MAXCOUNTS 1023.0
#elif ARDUINO_ARCH_ESP8266
  // On ESP8266:  0 - 1023 maps to 0 - 1 volts
  #define VOLTAGE_MAX 1.0
  #define VOLTAGE_MAXCOUNTS 1023.0
#endif
unsigned long myChannelNumber = 293695;
const char * myWriteAPIKey = "I0T24EFL1FSDKZEO"; //API key from
thingspeak
String TEMP_HUM_STRING = "";        // a string to hold incoming data
```

```
String SOIL_level,LIGHT_level,RAIN_level,FIRE_level,TEMP_
level,HUM_level,PRESS_level,ALT_level;
void setup()
{
Serial.begin(9600);
rajSerial.begin(9600);
#ifdef ARDUINO_AVR_YUN
Bridge.begin();
#else
#if defined(ARDUINO_ARCH_ESP8266) || defined(USE_WIFI101_
SHIELD) || defined(ARDUINO_SAMD_MKR1000)
WiFi.begin(ssid, pass);
#else
Ethernet.begin(mac);
#endif
#endif
ThingSpeak.begin(client);
}
void loop()
{   serialEvent_NODEMCU();
ThingSpeak.setField(1,SOIL_level);
ThingSpeak.setField(2,LIGHT_level);
ThingSpeak.setField(3,RAIN_level);
ThingSpeak.setField(4,FIRE_level);
ThingSpeak.setField(5,TEMP_level);
ThingSpeak.setField(6,HUM_level);
ThingSpeak.setField(7,PRESS_level);
ThingSpeak.setField(8,ALT_level);
delay(200);
```

```
Serial.print(SOIL_level);
Serial.print(";");
Serial.print(LIGHT_level);
Serial.print(";");
Serial.println(RAIN_level);
Serial.print(";");
Serial.print(FIRE_level);
Serial.print(";");
Serial.print(TEMP_level);
Serial.print(";");
Serial.print(HUM_level);
Serial.print(";");
Serial.print(PRESS_level);
Serial.print(";");
Serial.println(ALT_level);
#ifndef ARDUINO_ARCH_ESP8266
#endif
ThingSpeak.writeFields(myChannelNumber, myWriteAPIKey);
delay(20000); // delay20 sec
}
void serialEvent_NODEMCU()
{
while (rajSerial.available()>0)
{
TEMP_HUM_STRING = rajSerial.readStringUntil('\n');// Get serial input
StringSplitter *splitter = new StringSplitter(TEMP_HUM_STRING, ',', 8);
// new StringSplitter(string_to_split, delimiter, limit)
int itemCount = splitter->getItemCount();
for(int i = 0; i < itemCount; i++)
```

```
{
String item = splitter->getItemAtIndex(i);
SOIL_level = splitter->getItemAtIndex(0);
LIGHT_level = splitter->getItemAtIndex(1);
RAIN_level = splitter->getItemAtIndex(2);
FIRE_level= splitter->getItemAtIndex(3);
TEMP_level=splitter->getItemAtIndex(4);
HUM_level=splitter->getItemAtIndex(5);
PRESS_level=splitter->getItemAtIndex(6);
ALT_level=splitter->getItemAtIndex(7);
}
TEMP_HUM_STRING= "";
delay(20);
}
}
```

11.4 Main Server

Steps to Create a Channel

1. Sign In to ThingSpeak by creating a new MathWorks account.
2. Click Channels > MyChannels.

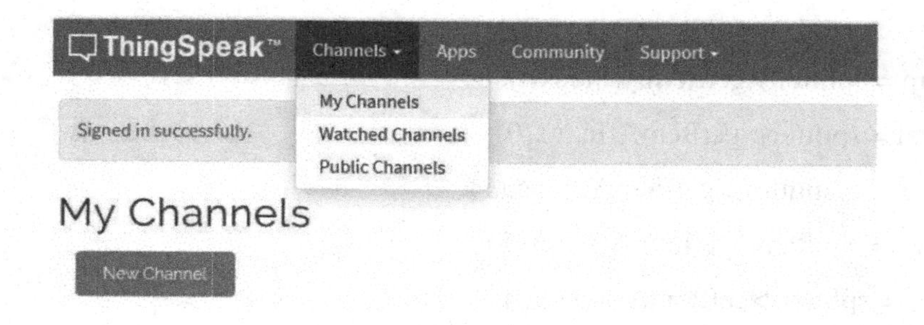

Fig.11.6: Window for thing speak

Name	Created	Updated At
Channel 293693 Private | Public | Settings | Sharing | API Keys | Data Import / Export	2017-06-26	2017-09-20 04:23

Fig.11.7: New channel in my channels

1. Click New Channel
2. Check the boxes next to Fields 1–1. Enter the channel setting values as follows:

Field 1: SOIL_level

Field 2: LIGHT_level

Field 3: RAIN_level

Field 4: FIRE_level

Field 5: TEMP_level

Field 6: HUM_level

Field 7: PRESSURE_level

Field 8: Altitude_level

1. Click Save Channel at the bottom of the settings
2. Check API write key (this key needs to write in the program for local server)

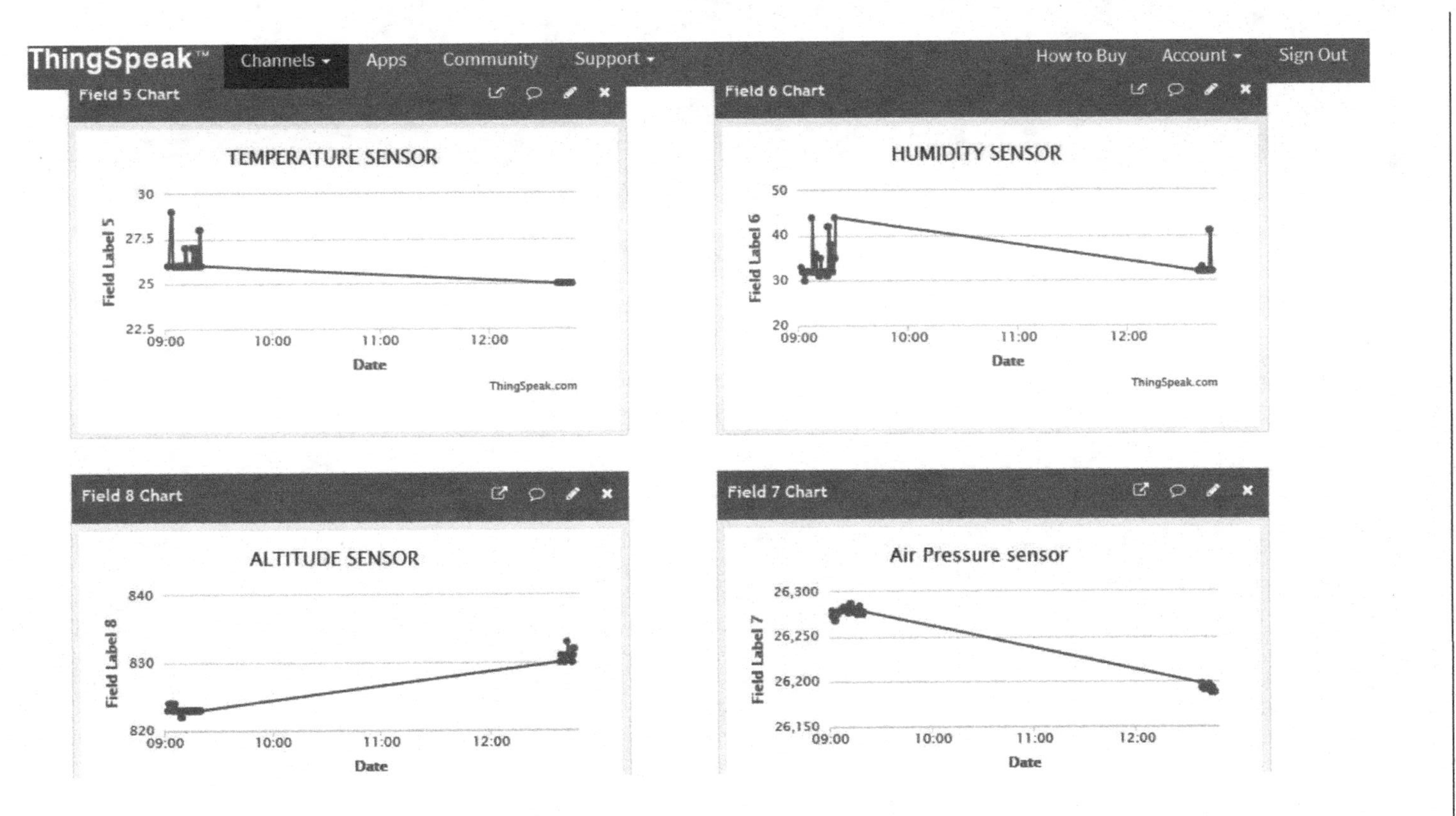

Fig.11.8: Charts from the field showing data received from local server

Printed in the United States
by Baker & Taylor Publisher Services